Comprehen

and Acronym Guide

Comprehensive Networking Glossary and Acronym Guide

GARY SCOTT MALKIN

MANNING
Greenwich

The publisher offers discounts on this book when ordered in quantity for special sales. For more information, please contact:

Special Sales Department
Manning Publications Co.
3 Lewis Street
Greenwich, CT 06830

Fax: (203) 661-9018
email: 73150.1431@compuserve.com

Recognizing the importance of preserving what has been written, it is the policy of Manning to have the books they publish printed on acid-free paper, and we exert our best efforts to that end.

Library of Congress Cataloging-in-Publication Data
Malkin, Gary Scott, 1961–
Comprehensive networking glossary and acronym guide / Gary Scott Malkin.
p. cm.
ISBN 1-884777-02-3 (acid-free paper)
1. Telecommunication systems—Terminology. 2. Telecommunication systems—Acronyms. I. Title
TK5102.M35 1994
004.6'014—dc20 94-19906
CIP

Manning Publications Co.
3 Lewis Street
Greenwich, CT 06830

Design: Frank Cunningham
Copyediting: Margaret Marynowski
Typesetting: Aaron Lyon

Printed in the United States of America
10 9 8 7 6 5 4 3

Preface

Greetings Gentlebeing!

My name is Gary Malkin. I was the principal editor of the *Internet Users' Glossary.* It is a very popular document within the Internet community. By design, that glossary was limited to Internet-specific terms, along with a few supporting terms. This means that it was very IP-centric. Using that glossary as a starting point, I have created the book you now hold.

The Comprehensive Networking Glossary and Acronym Guide (whew, quite a mouthful) contains 1501 terms and acronyms from AppleTalk, DECnet, IP, IPX, OSI, and telephony. It is richly cross-referenced, including three appendices listing Networks & Organizations, Security-Related Terms, and Applications & Protocols. There are also two dozen illustrations.

My electronic mail address is *gmalkin@xylogics.com.* If you who have access to the Internet, I hope, as you use this book, that you will take the time to send questions, suggestions, and, most importantly, corrections to me. Without your input, a second edition won't be as useful to you as you might like.

Gary Malkin

Acknowledgments

I used the *Internet Users' Glossary* (RFC 1392, FYI 18) as a starting point for this book, so I start my acknowledgments by thanking Tracy LaQuey Parker and all the people who contributed to that document. The entries in this book have been derived from numerous sources. Frequently it was necessary to merge multiple descriptions to form the final entries. Some of the entries were provided, on short notice, by Susan Calcari (InterNIC), David Conrad (APNIC), Martyne Hallgren (Cornell University), Judith Kiers (RARE), and Marten Terpstra (RIPE NCC).

I would also like to thank the people who helped review some of the entries, especially Jim DeMarco (FTP Software), Jon Postel (ISI), and Joyce Reynolds (ISI).

Finally, a very special thanks to Caroline Lange (Xylogics), who helped me convert this book from flat ASCII into FrameMaker.

Thanks Skippy, for getting me involved in User Services.

Introduction

They say the American English language grows faster than all other languages combined. One reason is the need to name each of the new technological innovations. Computer networking is one of the fastest growing technologies, and the volume of related terminology is growing proportionately fast. This book is written in such a way as to be useful to everyone from network neophyte to network developer.

The glossary entries are alphabetical within the body of the book. Hyphenated entries are sorted as though the hyphen were a blank; that is: "inter-domain" comes before "interactive." Also, acronym expansions are interspersed among the actual definitions. In general, an acronym expansion references the definition by saying "See..." In some cases, acronyms are defined via the acronym's expansion alone.

Several conventions are followed throughout the book.

- Commonly used acronyms for terms defined in the book are given in parentheses.
- Most acronyms appear entirely in UPPER CASE for ease of reading. Where an acronym is commonly expressed in mixed case, it will be so expressed in the book.
- If a term used within a definition is also defined in the book, it is indicated in *italic*. Only terms relevant to the term being defined are indicated, otherwise there would be a confusing number of cross-references.
- A "See also" following a definition indicates a list of other related definitions.

Some of the definitions refer to an RFC (look it up). RFCs may be obtained in many ways, depending on the type of Internet access available to you. They may also be obtained by mail. For more information, contact the InterNIC at +1 800 444 4345 or +1 619 455 4600.

Symbols

:-)

Symbols like this are used to introduce emotion into the flat medium of *Electronic Mail.* There are hundreds of such symbols, from the obvious to the obscure. This particular symbol is a smiley face, which expresses happiness or sarcasm. Don't see it? Tilt your head to the left 90 degrees.

@

The symbol used to separate the *username* from the *hostname* in an *email address.* See also: *Fully Qualified Domain Name.*

!

A symbol used to separate *hostnames* in an *email address.* See: *bang, bang path.*

10Base2

The formal designation for what is commonly referred to as Thinnet or Cheap(er)net. It refers to the physical *medium* which operates at 10Mb/s *baseband,* as opposed to *broadband,* and which has a maximum *segment* length (i.e., the maximum distance between *repeaters*) of 200 meters. See also: *broadcast medium, 802.3, Ethernet.*

10Base5

The formal designation for what is commonly referred to as Thicknet or, quite incorrectly, *Ethernet.* It refers to the physical *medium* which operates at 10Mb/s *baseband,* as opposed to *broadband,* and which has a maximum *segment* length (i.e., the maximum distance between *repeaters*) of 500 meters. See also: *broadcast medium, 802.3.*

10BaseT

The formal designation for what is commonly, and misleadingly, referred to as Twisted-Pair *Ethernet.* It refers to the physical *medium* which operates at 10Mb/s *baseband*, as opposed to *broadband*, and which operates over *Twisted-Pair* wire. See also: *broadcast medium, 802.3.*

1Base5

The formal designation for what is commonly referred to as StarLAN. It refers to the physical *medium* which operates at 1Mb/s *baseband*, as opposed to *broadband*, and which has a maximum *segment* length (i.e., the maximum distance between *repeaters*) of 500 meters. See also: *broadcast medium.*

23B+D

See: *Primary-rate ISDN*

2B+D

See: *Basic-rate ISDN*

802.2

The *IEEE* standard for the *Logical Link Control* sublayer of the *datalink layer.* It is used in conjunction with a *Media Access Control* sublayer, as defined in other *802.x* standards. See also: *802.3, 802.4, 802.5.*

802.3

The *IEEE* standard *Media Access Control* for *CSMA/CD.* It is used in conjunction with *802.2* to form a complete *datalink layer.* See also: *Ethernet, bus.*

802.4

An *IEEE* standard *Media Access Control* for *CSMA/CA* bus topologies. It is used in conjunction with *802.2* to form a complete *datalink layer.* See also: *token bus, bus.*

802.5

An *IEEE* standard *Media Access Control* for *CSMA/CA* ring topologies. It is used in conjunction with *802.2* to form a complete *datalink layer.* See also: *token ring.*

802.x

The set of *IEEE* standards for the definition of *Local Area Network* protocols. See also: *802.2, 802.3, 802.4, 802.5.*

822

See: *RFC 822*

A and B signaling

A form of *in-band signaling* used in *T1* transmissions. Bits are "stolen" from each of the *subchannels* and used to carry dialing and control information.

AARP

See: *AppleTalk Address Resolution Protocol*

ABM

See: *Asynchronous Balanced Mode*

abort sequence

A series of 12 to 18 1-bits appearing at the end of an AppleTalk *LLAP* frame. The sequence delineates the end of the *frame.*

Abstract Handle Specification (AHS)

The set of *DECmcc* functions which are used to establish and manipulate context *handles.*

abstract syntax

A description of a data structure that is independent of machine-oriented structures and encoding. See also: *Abstract Syntax Notation One, External Data Representation.*

Abstract Syntax Notation One (ASN.1)

The language used by the *OSI* protocols for describing *abstract syntax.* This language is also used to encode *SNMP* packets. ASN.1 is defined in *ISO* documents 8824.2 and 8825.2. See also: *Basic Encoding Rules, External Data Representation.*

Acceptable Use Policy (AUP)

Many *transit networks* have policies which restrict the use to which the network may be put. A well known example is *NSFNET*'s *AUP* which does not allow commercial use. Enforcement of AUPs varies with the network. See also: National Science Foundation.

Access Control List (ACL)

Most network security systems operate by allowing selective use of services. An Access Control List is the usual means by which access to, and denial of, services is controlled. It is simply a list of the services available, each with a list of the hosts permitted to use the service.

access line

The segment of a *leased line* which permanently connects the subscriber's *Point of Presence* to the *RBOC*.

access method

The mechanism or *protocol* by which a device or user may connect to or use a *network* or *service*.

Accunet

Data-oriented *digital* services offered by AT&T Communications. The 1.544-Mb/s services are primarily used for teleconferencing. Low-speed services are also available.

ACF

See: *Advanced Communications Function*

ACK

See: *Acknowledgment*

acknowledgment (ACK)

A type of *message* sent to indicate that a block of data arrived at its destination without error. See also: *Negative Acknowledgment.*

AC

See: *Access Control List*

acoustic coupler

A *modem* to which a telephone handset may be connected, rather than a modem which connects directly to a telephone *jack*. Data rates are limited to 1200b/s.

ACSE

See: *Association Control Service Element*

active device

A communications device which requires external power to operate. It typically offers more capabilities than a *passive device.* Most devices which operate above the *physical layer* are active devices.

active monitor

A *station* on an *802.5 LAN* which is responsible for ensuring that the *token* does not get lost, typically by line noise caused when a station enters or leaves the ring. The active monitor is not a special device, but a *peer* station on the network which has been "elected" to the task. It prevents other stations from attempting to become the active monitor by periodically transmitting an *Active Monitor Present* control *frame.*

Active Monitor Present (AMP)

A special *802.5* control *frame* which is periodically transmitted by the *active monitor.* If other stations on the ring do not hear an AMP for a period of time, they will "elect" a new active monitor.

AD

See: *Administrative Domain*

Adaptive Pulse Code Modulation (ADPCM)

A *CCITT* standard *encoding* technique which allows an *analog* voice conversion to be transmitted on a 32Kb/s *digital* communications channel. See also: *Pulse Code Modulation.*

address

A generic term for an identifier of a user, host, network or service. There are three types of addresses in common use within the *Internet.* They are *email address; network-layer address;* and *hardware address.* See also: *MAC address, Fully Qualified Domain Name.*

address mask

A bit mask used to identify which bits in an *internet address* correspond to the *network* and *subnet* portions of the address. This mask is often referred to as the subnet mask because the network portion of the address can be determined by the encoding inherent in an internet address.

address resolution

Conversion of an *internet address* into a corresponding *hardware address.*

Address Resolution Protocol (ARP)

Used to dynamically discover the low-level, *hardware address* which corresponds to the high-level *network-layer address* for a given host. ARP is limited to operation on a *broadcast medium.* See also: *proxy ARP.*

ADMD

See: *Administration Management Domain*

Administration Management Domain (ADMD)

An X.400 *Message Handling System* public service provider. Together, the global ADMDs form the *X.400* backbone. See also: *Private Management Domain.*

Administrative Domain (AD)

A collection of *hosts* and *routers,* and the interconnecting *network*(s), managed by a single administrative authority. See also: *Autonomous System.*

ADPCM

See: *Adaptive Pulse Code Modulation*

ADSP

See: *AppleTalk Data Stream Protocol*

Advanced Communications Facility (ACF)

IBM communications software product family which adds *SNA* functions to other, non-application systems' software.

Advanced Program-to-Program Communication (APPC)

An *SNA Application Program Interface* which allows computers to communicate without the intervention of a mainframe computer. APPC is also known as LU6.2

Advanced Research Projects Agency Network (ARPANET)

A pioneering *long-haul* network funded by *ARPA* (now *DARPA*). It served as the basis for early networking research, as well as a central *backbone* during the development of the *Internet.* The ARPANET consisted of individual packet switching computers interconnected by *leased lines.*

AEP

See: *AppleTalk Echo Protocol*

AES

See: *Application Environment Specification*

AFI

See: *AppleTalk Filing Interface, Authority and Format Identifier*

AFP

See: *AppleTalk Filing Protocol*

agent

In the *client-server model*, the part of the system which performs information preparation and exchange on behalf of a *client* or *server* application. In *network management*, the portion of an *entity* which responds to network management *queries*.

AHS

See: *Abstract Handle Specification*

AIX

Advanced Interactive Executive is a version of the *UNIX* operating system which operates on IBM RT workstations.

alias

A name, usually short and easy to remember, that is translated into another name, usually long and difficult to remember.

Aloha

The first **Media Access Control** method for *packet radio* networks. A node transmits whenever it has data to send, without first checking to see if another station is already transmitting. If an acknowledgment is not received within a certain period of time, the data is *retransmitted*. See also: *slotted Aloha*.

alternate routing

A feature of some *PBXs* wherein a call is completed over a second-choice *circuit* when the first-choice circuit is unavailable.

AM

See: *Amplitude Modulation*

American National Standards Institute (ANSI)

This non-profit, non-governmental organization is responsible for approving voluntary standards in the United States. It is composed of manufacturers, users and communications carriers, and handles standards for computer and communications technologies. ANSI is a voting member of *ISO*.

American Standard Code for Information Interchange (ASCII)

A standard character-to-number encoding widely used in thc computer industry. See also: *EBCDIC*.

Ameritech

The *Regional Bell Operating Company* which services the Mid-western region of the United States.

AMP

See: *Active Monitor Present*

Amplitude Modulation (AM)

A transmission method in which data is *encoded* over a *carrier* signal by variations in the voltage or current waveform. See also: *Frequency Modulation, Phase Modulation, Pulse Width Modulation*.

analog

A continuous form of transmission signal. Voice transmissions are analog. See also: *digital*.

annular mark

The mark on a *10Base5* cable which indicates the proper place to install a *transceiver*. The specification requires that distance between transceivers be a multiple of 2.5 meters.

anonymous FTP

A use of *FTP* which allows a user to retrieve documents, files, programs, and other data from anywhere in the *Internet* without having to establish a userid and password. By using the special userid of *anonymous* the network user will bypass local security checks and will have access to publicly accessible files on the remote system. See also: *archive site*.

ANSI

See: *American National Standards Institute*

answerback

The response of a communications device to a signal received from a remote device. Answerback is generally part of a two-way *handshake* which establishes a connection between devices.

AOW

See: *Asia and Oceania Workshop*

API

See: *Application Program Interface*

APNIC

The Asia Pacific Network Information Center, currently in a pilot project phase, is the delegated *Internet Registry* for Asia and the Pacific Rim. It also provides a forum for coordination of regional and national *Network Information Centers.*

APPC

See: *Advanced Program-to-Program Communication*

Appletalk

A network communications protocol developed by Apple Computer for communication between Apple Computer products.

AppleTalk Address Resolution Protocol (AARP)

The basic *Address Resolution Protocol* operated over *ELAP* and *TLAP.* It is used to determine a node's *hardware address* based on its *Appletalk* address.

AppleTalk Data Stream Protocol (ADSP)

An AppleTalk *session-layer* protocol which provides a *connection-oriented, stream-oriented* service between two *sockets.* See also: *AppleTalk Session Protocol, Printer Access Protocol.*

AppleTalk Echo Protocol (AEP)

A simple protocol which returns to the sender an exact copy of the packet it sent. It may be used to determine if a remote node is accessible. See also: *Packet Internet Groper.*

AppleTalk Filing Interface (AFI)

The *Application Program Interface* to the *AppleTalk Filing Protocol.*

AppleTalk Filing Protocol (AFP)

An AppleTalk *presentation-layer* protocol which allows users to access files and applications which reside on a remote node, typically a *file server.*

AppleTalk Remote Access Protocol (ARAP)

The AppleTalk *datalink-layer protocol* user over *serial interface* connections. See also: *EtherTalk Link Access Protocol, LocalTalk Link Access Protocol, Token-Talk Link Access Protocol.*

AppleTalk Session Protocol (ASP)

An AppleTalk *session-layer* protocol which provides session establishment, maintenance and takedown. See also: *AppleTalk Data Stream Protocol, Printer Access Protocol.*

AppleTalk Transaction Protocol (ATP)

The AppleTalk *transport-layer* protocol which provides a loss-free, transaction-based (*request/response*) service. Additional reliability (e.g., ordering) is available from *session-layer* protocols.

application

A program which performs a function directly for a user. *FTAM, Electronic Mail* and *Telnet* clients are examples of network applications.

Application Environment Specification (AES)

The *Open Software Foundation's* "look and feel" *standards.* They are based on the *X Windows* system.

application layer

The top (seventh) layer of the *OSI reference model.* This layer provides the interface to the user and is responsible for formatting user data before passing it to the lower layers for transmission to a remote host. *FTP, FTAM,* and *Electronic Mail* are examples of application protocols and functions.

Application Program Interface (API)

A set of calling conventions which define how a service is invoked through a software package. See also: *interface.*

ARAP

See: *AppleTalk Link Access Protocol*

archie

A system to automatically gather, index and provide information across the *Internet.* The initial implementation of archie provided an indexed directory of filenames from all *anonymous FTP* archives on the Internet. Later versions provide other collections of information. See also: *archive site, Gopher, Prospero, Wide Area Information Servers.*

architecture

The high-level principals which guide the design of a hardware or software system, product, or family of systems or products. An architecture specifies styles and methods, not nuts and bolts. See also: *hierarchical architecture, layered architecture, network architecture.*

archive site

A network node which provides access to a collection of files across the *Internet.* An *anonymous FTP* archive site, for example, provides access to this material via *FTP.* See also: *anonymous FTP, archie, Gopher, Prospero, Wide Area Information Servers.*

ARCnet

A 2.5Mb/s *Local Area Network,* developed by Datapoint and other companies. It uses a *star network topology.*

area

The *DECnet* term for *Autonomous System.* Level 1 *routers* operate strictly within one area; level 2 routers forward packets between areas. See also: *intra-domain router, inter-domain router.*

ARP

See: *Address Resolution Protocol*

ARPA

See: *Defense Advanced Research Projects Agency*

ARPANET

See: *Advanced Research Projects Agency Network*

ARQ

See: *Automatic Repeat Request*

ARS

See: *Automatic Route Selection*

AS

See: *Autonomous System*

ASCII

See: *American Standard Code for Information Interchange*

Asia and Oceania Workshop (AOW)

A regional *OSI* Implementors Workshop. It is a peer to the *European Workshop for Open Systems* and the *Workshop for Implementators of OSI.*

ASN.1

See: *Abstract Syntax Notation One*

ASP

See: *AppleTalk Session Protocol*

assigned numbers

The *RFC* which documents the currently assigned values from several series of numbers used in network protocol implementations. This RFC is updated periodically and, in any case, current information can be obtained from the *Internet Assigned Numbers Authority.* If you are developing a *protocol* or *application* which will require the assignment of a *port* number, protocol number, etc., please contact the *IANA* to receive a number assignment. See also: *STD.*

Association Control Service Element (ACSE)

An *OSI* data structure used to establish a call between two *application entities.*

asymmetric route

A condition in which the *path* from one *node* to another is not the same as the path from the second node to the first. While this situation is not common, neither is it inherently objectionable.

asynchronous

A transmission method which does not depend on the timing of the transmission facility. Asynchronous communications are character oriented and use *start bits* and *stop bits* to maintain device synchronization. See also: *synchronous.*

Asynchronous Balanced Mode (ABM)

An operational mode of IBM *token ring* networks. Within the **SNA** *datalink layer*, ABM allows a device to send datalink commands and initiate responses independently.

Asynchronous Transfer Mode (ATM)

A method for the dynamic allocation of *bandwidth* using a fixed-size packet (called a cell). ATM is also known as "fast packet."

ATM

See: *Asynchronous Transfer Mode*

ATP

See: *AppleTalk Transaction Protocol*

attenuation

Reduction in signal strength due to resistance in the transmission *medium* or the distance travelled by the signal or both. See also: *gain.*

attribute

A property, associated with an *entity*, which may be examined and modified by a *network management* protocol. For example, line speed is an attribute for communication *links.*

audit trail

A record of activity on a network or node. Audit trails may be generated to track access to a particular node or service for security reasons, or may be used to generate network utilization statistics. See also: *filtering, firewall.*

AUI

See: *Auxiliary Unit Interface*

AUP

See: *Acceptable Use Policy*

authentication

The verification of the identity of a person or process. See also: *authorization.*

Authority and Format Identifier (AFI)

The portion of an *OSI address* which indicates the type and format of the remaining portion of the address.

authorization

The determination of the access rights of a person or process. See also: *authentication.*

autobaud

The ability of a *modem* to recognize the *baud* rate of an incoming *call* and alter its own baud rate to match. The recognition is done by the hardware, not by a protocol.

Automatic Repeat Request (ARQ)

A request for *retransmission*, sent by the data recipient, to the data transmitter. See also: *go-back-N, selective-repeat, stop-and-wait.*

Automatic Route Selection (ARS)

The ability of a *PBX* to determine and use the optimum route when establishing a connection. ARS is also called Least Cost Routing.

Autonomous System (AS)

A collection of *routers* under a single administrative authority using a common *Interior Gateway Protocol* for exchanging routing information. See also: *Administrative Domain.*

Auxiliary Unit Interface (AUI)

The standard hardware *interface* used to connect a device to a *10base5 transceiver.* It uses a *DB-15* connector.

AWG

American Wire Gauge

B

B channel

A 64Kb/s *ISDN* data channel. See also: *Basic-rate ISDN, Primary-rate ISDN, D channel.*

B-ISDN

See: *Basic-rate ISDN*

backbone

The top level in a *hierarchical network. Stub networks* and *transit networks* which connect to the same backbone are guaranteed to be interconnected.

Backus-Naur Form (BNF)

An rigidly codified method for representing the elements (e.g., reserved words) and syntax of a computer language.

balun

In the IBM cabling system, balanced/unbalanced refers to an *impedance*-matching device used to connect balanced *Twisted-Pair* to unbalanced *coaxial cable.*

bandwidth

Technically, the difference, in *Hertz* (Hz), between the highest and lowest frequencies of a transmission channel. However, as typically used, the amount of data that can be sent through a given communications circuit.

bang

A single syllable reference to an exclamation point.

bang path

A series of machine names used to direct *Electronic Mail* from one user to another, typically by specifying an explicit *UUCP* path through which the mail is to be routed. See also: *bang, email address, mail path.*

baseband

A transmission *medium* through which *digital* signals are sent without complicated frequency shifting. In general, only one communication channel is available at any given time. See also: *XbaseY, broadband.*

BASIC

Beginners All-purpose Symbolic Instruction Code (computer programming language).

Basic Encoding Rules (BER)

Standard rules for *encoding* data units described in *ASN.1*. BER is sometimes incorrectly lumped under the term ASN.1, which properly refers only to the *abstract syntax* description language, not the encoding technique.

Basic-rate ISDN (B-ISDN)

An *ISDN* service which provides two 64Kb/s data/voice (B) channels plus a 16Kb/s control (D) channel. This service is also called 2B+D. See also: *Primary-rate ISDN.*

Basic Telecommunications Access Method (BTAM)

An early (pre-**SNA**) IBM control program which managed the remote data communications interface for applications.

batch

The processing of a user job which requires no interaction with the user. The results of the job are returned to the user once the job is complete. See also: *interactive.*

baud

A measurement of the signaling speed of a data communications device or network. It is equal to maximum number of *signal elements* which may be transmitted or received per second. A baud is not necessarily equal to the bits per second rating of a device or network because multiple bits may be *encoded* in one signal element using advanced encoding techniques.

Baudot code

An old data transmission code which uses five bits to represent a character.

Bayonet Nut Connector (BNC)

The standard connector used to connect a device to a *10base2* network.

BBS

See: *Bulletin Board System*

BCD

See: *Binary Coded Decimal*

BCNU

Be Seein' You

beacon

An *802.5* control *frame* which indicates a serious failure on the ring. The beacons are sent by whichever *stations* detect the failure.

Bell Atlantic

The *Regional Bell Operating Company* which services the Mid-atlantic region of the United States.

Bell South

The *Regional Bell Operating Company* which services the southeastern region of the United States.

Bellcore

The Bell Communications Research organization was established by the AT&T divestiture. It represents, and is funded by, the *RBOCs* and establishes the *telephony* standards and *interfaces.*

BER

See: *Basic Encoding Rules*

Berkeley Internet Name Domain (BIND)

An implementation of a *Domain Name System* server developed and distributed by the University of California at Berkeley. Many *Internet* hosts run BIND, and it is the ancestor of many commercial BIND implementations.

Berkeley Software Distribution (BSD)

An implementation of the *UNIX* operating system and its utilities developed and distributed by the University of California at Berkeley. "BSD" is usually preceded by the version number of the distribution. For example, "4.3 BSD" is version 4.3 of the Berkeley UNIX distribution. Many *Internet* hosts run BSD software, and it is the ancestor of many commercial UNIX implementations.

BGP

See: *Border Gateway Protocol*

big-endian

A format for storage or transmission of binary data in which the *Most Significant Bit/Byte* comes first. The term comes from "Gulliver's Travels" by Jonathan Swift. The Lilliputians, being very small, had correspondingly small political problems. The Big-Endian and Little- Endian parties debated over whether soft-boiled eggs should be opened at the big end or the little end. See also: *byte order, little-endian.*

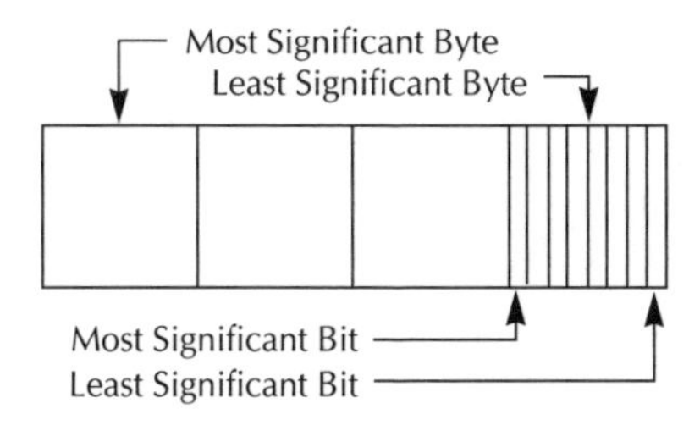

BIH

See: *International Time Bureau*

binary

The base-2 numbering system used by digital computers. See also: *bit, octal, hexadecimal.*

Binary Coded Decimal (BCD)

An old numeric code wherein the numbers zero through nine each have a unique 4-bit binary representation.

Binary Synchronous Communications (BSC)

An IBM, *character-oriented* data communications protocol. While still widely deployed, it is being replaced by *SDLC.*

BIND

See: *Berkeley Internet Name Domain*

bipolar

The most common signaling method used in *digital* transmission services. The zero and one values are represented by different signal voltage levels, and each bit is the opposite polarity of the preceding bit. See also: *electrical signaling* (includes figure).

Birds Of a Feather (BOF)

A Birds Of a Feather (flocking together) is an informal discussion group. It is formed, often ad hoc, to consider a specific issue and, therefore, has a narrow focus. See also: *Internet Engineering Task Force, Working Group.*

bisync

See: *Binary Synchronous Communications*

bit

A binary digit is the smallest unit of information available, having only two values (zero and one). See also: *binary, signal element.*

bit duration

The amount of time it takes for one *encoded* bit to pass a point on a transmission *medium.* Bit duration is a function of the transmission rate a specific transmission medium. For example, a 1Mb/s transmission medium has a bit duration of one-millionth of a second.

bit-oriented

A communications protocol in which information is encoded in fields of bits, as opposed to bytes. *HDLC* and *SDLC* are examples of bit-oriented protocols. See also: *character-oriented.*

bit stuffing

Used in *bit-oriented* communications protocols, bit stuffing is the process of inserting zero-bits into the data stream in order to prevent user data from being interpreted as protocol control information (usually a *flag*). These stuffed bits are removed by the receiver to restore the data before it is handed up to the user.

Bitnet

An academic computer network which provides *Electronic Mail* and *file transfer* services, using a *store-and-forward* protocol Bitnet is based on IBM Network Job Entry protocols. Bitnet-II *encapsulates* the Bitnet protocol within *IP* packets and depends on the *Internet* to route them.

black hole

A situation, typically caused by an erroneous *route* or a dead *router*, in which transmitted packets disappear. *Routing* and *forwarding* problems for which error *messages* are generated are not considered black holes, because the packet's originating node has some indication that something went wrong.

block multiplexor channel

An IBM mainframe input/output channel which permits the *interleaving* of blocks of data. See also: *byte multiplexor channel.*

BNC

See: *Bayonet Nut Connector*

BNF

See: *Backus-Naur Form*

BOC

Bell Operating Company (see: *Regional Bell Operating Company*)

BOF

See: *Birds Of a Feather*

bogon

A humorous term applied to a malformed or mis-addressed packet. Bogon is short for "bogus packet." See also: *Martian, kiss-of-death packet.*

BOOTP

The Bootstrap Protocol is used for booting *diskless nodes*. See also: *Dynamic Host Configuration Protocol, Reverse Address Resolution Protocol.*

Border Gateway Protocol (BGP)

An *Internet* Standard *Exterior Gateway Protocol.* Its design is based on experience gained with *EGPs*, usage in the *NSFNET.*

border router

See: *inter-domain router*

bounce

The return of a piece of *Electronic Mail* because of an error in its delivery.

boundary node

An *SNA* subarea node which can provide protocol support for adjacent subarea nodes. This support may include mapping *network addresses* and local addresses, *session-layer* sequencing, and *flow control.*

bridge

A device which *forwards* traffic between network segments based on *datalink-layer* information. These segments would have a common *network-layer address.* See also: *learning bridge, gateway, router.*

broadband

A transmission *medium* capable of supporting a wide range of frequencies. It can carry multiple signals by dividing the total *carrying capacity* of the medium into multiple, independent *bandwidth* channels, where each *channel* operates only on a specific range of frequencies. See also: *XbroadY, baseband, guard frequency.*

broadcast

A special type of *multicast* packet which all nodes on the network are always willing to receive. See also: *directed broadcast, network broadcast, subnet broadcast.*

broadcast medium

Any transmission *medium* which support the delivery of a single packet to multiple nodes attached to that medium. See also: *CSMA/CA, CSMA/CD*

broadcast storm

A *Martian* packet, broadcast onto a network, which causes multiple hosts to respond all at once, typically with equally incorrect packets which causes the storm to grow exponentially in severity.

brouter

A device which *bridges* some packets (i.e., *forwards* based on *datalink-layer* information) and *routes* other packets (i.e., forwards based on *network-*

layer information). The bridge/route decision is typically based on configuration information. See also: *bridge, router.*

BSC

See: *Binary Synchronous Communications*

BSD

See: *Berkeley Software Distribution*

BTW

By The Way

buffering

In a *store-and-forward* network device, incoming packets are held in memory to facilitate error checking and compensation for line speed and *MTU* mismatches.

bullet-proof

Software implementations which are impervious to kiss-of-death packets.

Bulletin Board System (BBS)

A computer, and associated software, which typically provides electronic messaging services, archives of files, and any other services or activities of interest to the bulletin board system's operator. Although BBSs have traditionally been the domain of hobbyists, an increasing number of BBSs are connected directly to the *Internet*, and many BBSs are currently operated by government, educational, and research institutions. See also: *Electronic Mail, Usenet.*

bus

A shared communication link onto which multiple nodes may *tap*. In *802.3*, for example, the tap is a *transceiver*, which is connected to the node by a transceiver cable. See also: *Ethernet, 802.4, drop cable.*

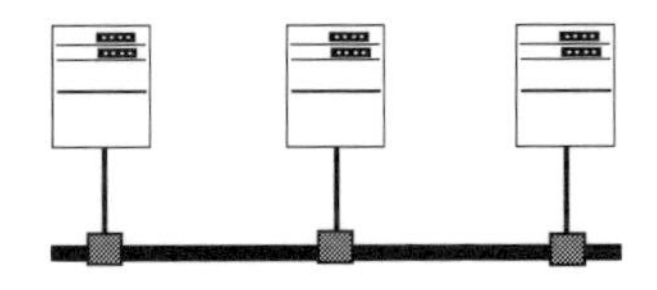

bus and tag

See: *OEM-I*

byte

A unit of information. It is usually eight *bits* in size and 8-bit aligned. See also: *octet, nibble, binary, octal, hexadecimal.*

byte multiplexor channel

An IBM mainframe input/output channel which permits the *interleaving* of bytes of data. See also: *block multiplexor channel.*

byte order

The order in which bytes in a word are stored in memory or transmitted over a network. See also: *host byte order, network byte order, big-endian, little-endian.*

byte-swap

The changing of data from *network byte order*, typically *big-endian*, to *host byte order.* In software which operates on multiple CPUs, 2-byte and 4-byte fields are passed to functions which return the values unchanged, if the host byte order matches the network byte order, or reversed, if the host and network byte orders are different. See also: *little-endian.*

C

C band

The portion of the *electromagnetic spectrum*, with frequencies ranging from 4GHz to 6GHz, used for microwave and satellite transmission.

CA

See: *Certification Authority*

CAD

Computer Aided Design

CAE

Computer Aided Engineering

call

An established *dial-up* connection. It may refer to a connection over the telephone system or over a *Public Data Network*.

call-forwarding

A feature offered by some *RBOCs* and *PBXs* which allows a subscriber to have incoming calls automatically transferred to another telephone number or PBX extension.

call-waiting

A feature offered by some *RBOCs* and *PBXs* which signals a subscriber, already engaged in one call, that another call is waiting, rather than simply signaling the second caller with a busy signal.

CAM

Computer Aided Manufacturing

CAM

See: *Content Addressable Memory*

camp-on

A feature offered on some *PBXs* which signals a user when a requested extension, previously busy, becomes available.

Campus Wide Information System (CWIS)

A system which makes information and services available, over a campus network, via *interactive*, public terminals located throughout the campus. Services typically include directory information, calendars, *Bulletin Board Systems*, and databases.

carrier

A continuous transmission signal, frequently a sine wave, which may *modulated* with a data signal. See also: *Amplitude Modulation, Frequency Modulation, Phase Modulation, Pulse Width Modulation.*

Carrier Detect (CD)

A *modem* signal which indicates that the remote modem has answered a call or the local modem has accepted a call. See also: *RS-232.*

Carrier Sense Multiple Access with Collision Avoidance (CSMA/CA)

A method for handling *contention* on shared media. "Carrier Sense" indicates that a station on the network will listen to the wire before trying to send, in order to capture the *token* which grants permission to send. "Multiple Access" indicates that many stations may be connected to the same wire as *peers*. "Collision Avoidance" indicates that the access method prevents multiple stations from colliding during transmit, usually by token acquisition. See also: token bus, token ring, CSMA/CD.

Carrier Sense Multiple Access with Collision Detection (CSMA/CD)

A method for handling *contention* on shared media. "Carrier Sense" indicates that a station on the network will listen to the wire before trying to send, to prevent a *collision* with another station that is currently transmitting. "Multiple Access" indicates that many stations may be connected to the same wire as *peers*. "Collision Detection" indicates that the access method used detects when two stations' transmissions are colliding and requires them both to stop immediately and try again after a random backoff. See also: *802.3, Ethernet, CSMA/CA.*

carrying capacity

The maximum *offered load* which a network is capable of handling. See also: *throughput, congestion.*

Catalog Node (CNode)

A directory or file entry in a disk's volume catalog. The *AppleTalk Filing Protocol* recognizes internal CNodes, represented by directories, and leaf CNodes, represented by empty directories of files.

catenet

An *internet* in which hosts are connected to networks with various different characteristics. These networks are interconnected with *routers*, capable of handling multiple protocols. The *Internet* is an example of a catenet.

Cathode Ray Tube (CRT)

Technically, the phosphor-coated vacuum tube with which images may be displayed. Colloquially, the device in which said vacuum tube is placed. See also: *teletypewriter.*

CATV

See: *Community Antenna Television*

CBX

See: *Computerized Branch Exchange*

CCIRN

See: *Coordinating Committee for Intercontinental Research Networks*

CCIS

See: *Common Channel Interoffice Signaling*

CCITT

See: *Comité Consultatif International Télégraphique et Téléphonique*

CCR

See: *Commitment, Concurrency and Recovery*

ccs

A unit of telecommunications traffic measurement (100 call seconds). See also: *Erlang.*

CD

See: *Carrier Detect*

cell

In *cellular radio*, the geographic area covered by a single transmitter. In *Asynchronous Transfer Mode*, a fixed-size packet.

cellular radio

A technology which uses low-power radio to replace the wire connection to the *Central Office*. A cellular phone may be stationary or mobile, in which case a handoff between cells occurs, transparently to the user, as cell boundaries are crossed.

Central Office (CO)

The telephone company switching office to which subscribers are attached via a *local loop*. A CO is identified by the first three digits of a seven-digit telephone number. A CO is also called an exchange.

CERT

See: *Computer Emergency Response Team*

Certification Authority

A software *process* or database which maintains *public keys* for individuals or *processes* which use *RSA*. See also: *encryption*.

Challenge Handshake Authentication Protocol (CHAP)

A *PPP* authentication protocol which uses a *three-way handshake* and *encrypted*, password-derived tokens to *authenticate* the remote side of the connection. See also: *Packet Authentication Protocol*.

channel

A physical or logical, voice or data communications *path* which connects two communications devices. See also: *block multiplexer channel, byte multiplexer channel, T1, T3*.

channel bank

Telephone *Central Office* equipment which multiplexes lower-speed lines, usually from multiple subscribers, into high-speed *channels* which interconnect the *CO*s.

Channel Service Unit (CSU)

A data communications device which terminates a *digital* circuit at the customer site. It is responsible for ensuring that all subscriber data complies with the service provider's and *FCC*'s rules regarding data transmission over the circuit to which it is attached. See also: *Data Service Unit, Customer Premises Equipment.*

CHAP

See: *Challenge Handshake Authentication Protocol*

character

A generic name for a small unit of information. Typically, a character is the same as a *byte* or an *octet*. See also: *control characters, printable characters.*

character-oriented

A communications protocol in which information is *encoded* in fields of *characters*, as opposed to *bits*. See also: *bit-oriented, control characters.*

cheap(er)net

See: *10base2*

checksum

A computed value which is dependent upon the contents of a packet. This value is sent along with the packet when it is transmitted. The receiving system computes a new checksum based upon the received data and compares this value with the one sent with the packet. If the two values are the same, the receiver has a high degree of confidence that the data was received correctly. See also: *Cyclic Redundancy Check, Frame Check Sequence.*

chokepoint

The single path between two *networks*, typically a *router*. Chokepoints are used to manage the flow of information onto and off of a network, because the more paths there are the harder it is to manage network access. See also: *firewall.*

CICS

See: *Customer Information Control System*

CIDR

See: *Classless Inter-Domain Routing*

CIM

Computer Integrated Manufacturing

circuit

A generic term for a physical or logical connection between two communications devices.

Circuit-Switched Digital Capability (CSDC)

A service, designed by AT&T and implemented within the *RBOCs*, which provides a 56Kb/s digital circuit to subscribers on a subscriber-switchable basis. The service uses the same *local loop* as a voice circuit. See also: *Public Switched Digital Service.*

circuit switching

A communications paradigm in which a dedicated communications *path* is established between two hosts, and on which all packets travel. The telephone system is an example of a circuit switched network. See also: *connection-oriented, connectionless, packet switching.*

CIT

See: *Computer Integrated Telephony*

CIX

See: *Commercial Information Exchange*

Classless Inter-Domain Routing (CIDR)

A strategy developed within the *IETF* to extend the usable lifetime of *IP* version 4 as the network layer for the *Internet.* It eliminates the notion of address class when using an *internet address* for routing. It is defined in RFC 1519.

Clear To Send (CTS)

A modem control signal which the *DCE* asserts to the *DTE* indicating that the DCE is ready to receive data. See also: *flow control, RS-232.*

clearinghouse

A database which may be queried to supply information within a specified *domain.* For example, a *DNS* server is a clearinghouse for *Fully Qualified Domain Names.*

Clearinghouse for Networked Information Discovery and Retrieval (CNIDR)

An organization who's mission is to promote and support the implementation and use of networked information discovery and retrieval software *applications,* coordinate to create consensus among NIDR applications developers to ensure compatibility and *interoperability*, and to disseminate information about NIDR applications to the network community. See also: *archie, Gopher, Prospero, Wide Area Information Server, World Wide Web.*

client

A computer system or process which requests a service of another computer system or process. A workstation requesting the contents of a file from a *file server* is a client of the file server. See also: *client-server model, server.*

client-server model

A common way to describe the paradigm of many network protocols. Examples include the name-server/name-resolver relationship in *DNS* and the file-server/file-client relationship in *NFS*. See also: *client, server.*

CLNP

See: *Connectionless Network Protocol*

closed user group

A group of users assigned to a network facility which may communicate with each other but not with users in other groups. *X.25* offers support of closed user groups.

CLTP

See: *Connectionless Transport Protocol*

cluster controller

A dedicated device which performs data communications functions for one or more terminals or workstations which cannot be connected directly to a network.

CMIP

See: *Common Management Information Protocol*

CMIP Over TCP (CMOT)

A version of *CMIP* which operates over *TCP* rather than *TP4*. It is not in common use.

CMIS

See: *Common Management Information Services*

CMOT

See: *CMIP Over TCP*

CNI

See: *Coalition for Networked Information*

CNIDR

See: *Clearinghouse for Networked Information Discovery and Retrieval*

CNode

See: *Catalog Node*

CNRI

See: *Corporation for National Research Initiatives*

CO

See: *Central Office*

Coalition for Networked Information (CNI)

A consortium formed by American Research Libraries, CAUSE, and EDUCOM to promote the creation of, and access to, information resources in networked environments in order to enrich scholarship and enhance intellectual productivity.

coaxial cable (coax)

A transmission *medium* consisting of a central, solid conductor (usually copper), surrounded by a dielectric insulator, surrounded by a wire mesh shield, surrounded by an outer insulating cover. A coaxial cable can typically support transmission frequencies between 50MHz and 500MHz. See also: *twinaxial cable, terminated line.*

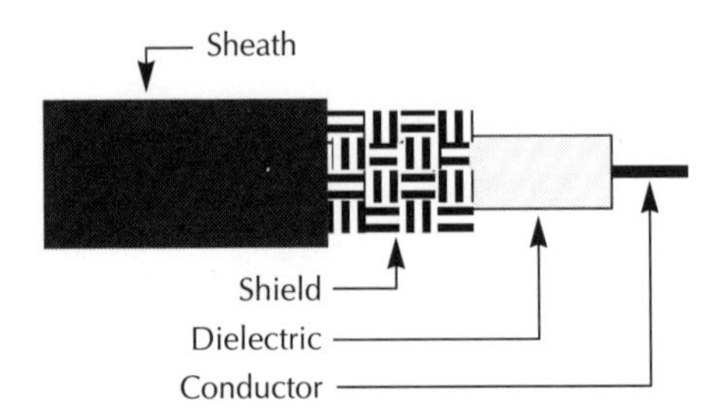

COBOL

Common Oriented Business Language (early computer programming language)

codec

See: *coder/decoder*

coder/decoder

A device which permits *analog* data (e.g., voice) to be transmitted over a *digital* communications network. See also: *modulator/demodulator.*

collision

A condition in which multiple nodes are transmitting on the same *medium* at the same time. The result is that all of the packets are damaged. The recovery method depends on the *Media Access Control* protocol being used. See also: *CSMA/CD.*

combined station

A communications device capable of acting as a primary or a secondary *HDLC* station.

Comité Consultatif International Télégraphique et Téléphonique (CCITT)

This organization is part of the *International Telecommunications Union* and is responsible for making technical recommendations about telephone and data communications systems. Every four years CCITT holds plenary sessions where they adopt new standards; the most recent was in 1992. Recently, the *ITU* reorganized and CCITT was renamed the *ITU-TSS.*

Command Terminal Protocol (CTERM)

A portion of the *Virtual Terminal* service in the *presentation layer* of the *Digital Network Architecture.* See also: *Local Area Transport.*

Commercial Information Exchange (CIX)

One of the connection points between the commercial carriers (e.g., Alternet, PSInet) serving the *Internet.* The commercial carrier networks are typically *backbone* networks, like *NSFNET.*

Commercial Off-The-Shelf (COTS)

A description of network components (e.g., *hosts*, *routers*, cables) which are usable as they are sold. That is, no custom hardware or software is required.

Commitment, Concurrency and Recovery (CCR)

An *OSI* application service element which executes an atomic (i.e., indivisible) operation across a distributed system.

Common Channel Interoffice Signaling (CCIS)

An AT&T method of separate-channel signaling in which control and signaling for a group of communication *trunks* between telephone *Central Offices* is carried in a separate *channel*. See also: *out-of-band signaling*.

common carrier

In the United States, a supplier of communications services or facilities which is authorized to provide those services and facilities by the appropriate government regulatory agency. Telephone and cable providers are examples of common carriers.

Common Management Information Protocol (CMIP)

The *OSI* standard *network management* protocol. See also: Management Information Base, Specific Management Functional Area, Simple Network Management Protocol.

Common Management Information Services (CMIS)

The *OSI* standard *network management* services which are provided by *CMIP*.

Community Antenna Television (CATV)

Uni-directional data communications based on *Radio Frequency* transmissions over 75-ohm *coaxial cable*. See also: *broadband*.

compression

A fully reversible reduction in the number of bits which need to be transmitted or stored in order to save *bandwidth* or storage, respectively. There are many compression algorithms available.

Computer Emergency Response Team (CERT)

The CERT was formed by *DARPA* in November 1988 in response to the needs exhibited during the Internet *worm* incident. The CERT charter is to work with the *Internet* community to facilitate its response to computer security events involving Internet hosts, to take proactive steps to raise the community's awareness of computer security issues, and to conduct research targeted at improving the security of existing systems. CERT products and services include 24-hour technical assistance for responding to computer security incidents, product vulnerability assistance, technical documents, and tutorials. In addition, the team maintains a number of mailing lists (including one for CERT Advisories), and provides an *anonymous FTP* server, at *cert.org*, where security-related documents and tools are archived. The CERT may be reached by *email* at *cert@cert.org* and by telephone at +1 412 268 0090 (24-hour hotline).

Computer Integrated Telephony (CIT)

Digital Network Architecture products which are used to integrate *PBXs* into *DNA* computer networks.

COMSAT

The Communications Satellite Corporation, a private United States satellite carrier established by Congress in 1962, constructs and coordinates *satellite communication* facilities used for international voice and data communications.

concentrator

An active communications device which provides *taps* for multiple network nodes on a shared communications medium. See also: *wiring hub.*

conditioned line

A dedicated telephone circuit on which the line *impedances* have been carefully balanced by the circuit provider (usually the telephone company). This special conditioning allows higher data transmission rates. Generally, several levels of conditioning are available at increasing additional costs. See also: *leased line.*

confidence

A *DECnet transport-layer* variable which indicates the probability that an open connection is still functioning.

conformance test

An empirical examination of a hardware or software product to determine how closely it corresponds to the specification governing its operation. See also: *interoperability.*

congestion

A condition which exists when the *offered load* exceeds the *carrying capacity* of a data communications network.

congestion control

A mechanism with which the *offered load* on a network is regulated so as not to exceed a network's *carrying capacity.* See also: *flow control.*

connection-oriented

The data communications method in which communication proceeds through three well-defined phases: connection establishment, data transfer, connection release. *TCP* and *TP4* are examples of connection-oriented protocols. See also: *circuit switching, connectionless, packet switching.*

Connection Oriented Network Service (CONS)

A light-weight protocol which allows *OSI transport-layer* protocols to bypass *CLNP* when operating over a single, logical *X.25* network.

connectionless

A data communications method in which communication occurs between hosts with no previous setup. Packets between two hosts may take different *routes,* as each is independent of the other. *UDP* is an example of a connectionless protocol. See also: *circuit switching, connection-oriented, packet switching.*

Connectionless Network Protocol (CLNP)

An *OSI network-layer* protocol which provides a *datagram* service similar to *IP.* CLNP is mandated by *GOSIP.*

Computerized Branch Exchange (CBX)

An electronic (as opposed to mechanical) *PBX.* Most PBXs manufactured today are CBXs. See also: *Private Automatic Branch Exchange.*

Connectionless Transport Protocol (CLTP)

An *OSI transport-layer* protocol which provides *multiplexing* and guarantees data integrity, but does not provide reliable delivery or *flow control.* It is similar to *UDP.*

CONS

See: *Connection Oriented Network Service*

Content Addressable Memory (CAM)

A special type of memory which returns an indication of whether or not a particular value has been stored. It is used by network devices which learn the *hardware addresses* of other devices on a shared network medium. See also: *learning bridge.*

contention

A situation which arises on shared resources in which multiple data sources compete for access to the resource. See also: *CSMA/CD.*

Continuous Variable-Slope Delta modulation (CVSD)

A speech *encoding* and *digitizing* technique which uses a one bit sample to encode the difference between two successive signals. The typical sampling rate is 32,000 samples per second.

control characters

Special *characters* inserted into a data stream which are used by the data communications devices to communicate and coordinate with each other. Control characters are used for framing, blocking, synchronization, and other special functions. See also: *EOA, EOB, EOM, EOT, ETB, ETX, SOH, SOM, STX, printable characters.*

convergence time

The time between a network *topology* change and the time all of the *routers* have up-to-date routing information. It is beneficial to have this time be as short as possible to prevent data packets from becoming lost. See also: *routing protocol.*

Cooperation for Open Systems Interconnection Networking in Europe (COSINE)

A European Commission sponsored program whose goal is to interconnect European research networks using *OSI.*

Coordinated Universal Time (UTC)

See: *Greenwich Mean Time*

coordinates

A generic term referring to how a person may be reached. It may include an *email address,* postal address, telephone or *facsimile* number, pager number, etc., or any combination of the above.

Coordinating Committee for Intercontinental Research Networks (CCIRN)

A committee that includes the United States *Federal Networking Council* and its counterparts in North America and Europe. Co-chaired by the executive directors of the *FNC* and *RARE*, the CCIRN provides a forum for cooperative planning among the principal North American and European research networking bodies.

core gateway

Historically, one of a set of *gateways* (*routers*) operated by the Internet *Network Operations Center* at Bolt, Beranek and Newman (BBN). The core gateway system formed a central part of *Internet* routing in that all groups must advertise *routes* to their networks from a core gateway.

Corporation for National Research Initiatives (CNRI)

A non-profit research and development organization formed in 1986 to help focus United States strengths in information processing technology. CNRI works with academia, industry and government in scientific research on the design of experimental infrastructure which can improve long-range scientific and engineering productivity. The *IETF* Secretariat, which manages IETF meetings and mailing lists, is composed of CNRI staff.

COS

See: *Corporation for Open Systems*

COSINE

See: *Cooperation for Open Systems Interconnection Networking in Europe*

Corporation for Open Systems (COS)

A vendors and users group which performs conformance testing of, certification for, and promotion of *OSI* products.

Corporation for Research and Educational Networking (CREN)

This organization was formed in October 1989, when *Bitnet* and CSNET (Computer + Science NETwork) were combined under one administrative authority. CSNET is no longer operational, but CREN still runs Bitnet.

COTS

See: *Commercial Off-The-Shelf*

CPE
See: *Customer Premises Equipment*

CPU
Central Processing Unit

cracker
An individual who attempts to access computer systems without *authorization*. These individuals are often malicious, as opposed to *hackers*, and have many means at their disposal for breaking into a system. See also: *Computer Emergency Response Team, Trojan Horse, virus, worm.*

CRC
See: *Cyclic Redundancy Check*

credit
A *LAT flow control* mechanism. A *node* may send only as many packets as it has credits. If it has packets to send and no credits, it must wait for the remote node to send more credits.

CREN
See: *Corporation for Research and Educational Networking*

crosstalk
The unwanted leakage of a communications signal from one electrical transmission medium to an adjacent medium. Typically, this problem occurs with *Unshielded Twisted-Pair* on voice and data lines.

CRT
See: *Cathode Ray Tube*

CSCD
See: *Circuit-Switched Digital Capability*

CSMA/CA
See: *Carrier Sense Multiple Access with Collision Avoidance*

CSMA/CD
See: *Carrier Sense Multiple Access with Collision Detection*

CSU
See: *Channel Service Unit*

CTERM

See: *Command Terminal Protocol*

CTS

See: *Clear To Send*

Customer Information Control System (CICS)

An IBM product and mainframe environment which enables transactions to be entered on remote terminals and concurrently processed by user applications.

Customer Premises Equipment (CPE)

Telephone equipment which resides at the subscribers site. See also: *Data Service Unit, Network Channel Terminating Equipment, Point Of Presence.*

CVSD

See: *Continuous Variable-Slope Delta modulation*

CWIS

See: *Campus Wide Information system*

Cyberspace

A term coined by William Gibson in his fantasy novel *Neuromancer* to describe the "world" of computers and the society which gathers around them.

Cyclic Redundancy Check (CRC)

A number derived from a set of data which will be transmitted. By recalculating the CRC at the remote end and comparing it to the value originally transmitted, the receiving node can detect many types of noise-related transmission errors.

D

D-bit

The *X.25* delivery confirmation bit. It indicates that the *DTE* wants an *end-to-end* acknowledgment of data delivery.

D channel

A 16Kb/s *B-ISDN* control channel, or a 64Kb/s *P-ISDN* control channel. See also: *out-of-band signaling, B channel.*

D4 framing

T1 12-bit frame format which uses the 193rd bit for framing and signaling. See also: *Extended Superframe Format.*

DACS

See: *Digital Access and Cross-connect System*

DAP

See: *Directory Access Protocol*

DAR

See: *Dynamic Adaptive Routing*

DARPA

See: *Defense Advanced Research Projects Agency*

DAS

See: *Dynamically Assigned Socket*

DASD

Direct Access Storage Device (often pronounced "daz-dee")

Data Circuit-terminating Equipment (DCE)

A data communications device which serves as an access point to a data communications network. A *modem* is an example of a DCE. DCE is often, erroneously, referred to as Data Communications Equipment. See also: *Data Terminal Equipment.*

Data Encryption Key (DEK)

Used for the *encryption* of *message* text and for the computation of message integrity checks (signatures). See also: *encryption.*

Data Encryption Standard (DES)

The standard cryptographic algorithm, designed by the *National Bureau of Standards*, used to *encrypt* and *decrypt* information using a 64-bit key. The algorithm is specified in *FIPS* Publication 46 (January 15, 1977). See also: *RSA.*

Data Network Identification Code (DNIC)

The four-digit number assigned to *Public Data Networks* and to specific services within those networks. DNICs serve the function of *addresses* and *SAP*s.

Data Service Unit (DSU)

A data communications device which interfaces to a *CSU* and is responsible for converting subscriber data into *bipolar* format.

Data Set Ready (DSR)

A *modem* signal which indicates that the modem is connected to a telephone line and is ready to send data to the remote modem. See also: *Data Terminal Ready, RS-232.*

Data Terminal Equipment (DTE)

Data communication devices, typically terminals and *Personal Computers*, which access data communication networks via *DCE*s.

Data Terminal Ready (DTR)

A *modem* signal which indicates to the modem that *DTE* is ready to communicate. See also: *Data Set Ready, RS-232.*

data transparency

See: *transparency*

datagram

A self-contained, independent entity of data carrying sufficient information to be *route*d from the source to the destination computer without reliance on earlier exchanges between this source and destination computer and the transporting network. The term was coined by Jon Postel, one of the fathers of the *Internet.* See also: *frame, packet.*

Datagram Delivery Protocol (DDP)

The AppleTalk *network-layer* protocol. It is a *connectionless,* best-cffort protocol similar to *IP* and *CLNP.*

datalink layer

The second layer of the *OSI reference model.* This layer is responsible for *hop-to-hop,* as opposed to *end-to-end,* delivery of data. *PPP* and *802.3* are examples of datalink-layer protocols.

Dataphone Digital Service (DDS)

A private line, digital service provided by *RBOCs* (*intraLATA*) and AT&T Communications (*interLATA*).

DB-9

A standard 9-pin connector used for *serial interfaces.* The diagram below shows a male connector looking at the pins. See also: *RS-232.*

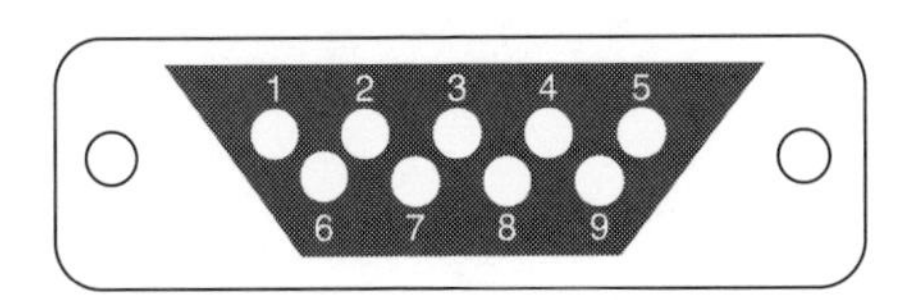

DB-15

A standard connector similar to a *DB-9,* but with 15 pins. This connector is specified for the *AUI* interface. See also: *10Base5.*

DB-25

A standard connector similar to a *DB-9,* but with 25 pins. This connector is specified for the *RS-232* interface.

DB-37

A standard connector similar to a *DB-9,* but with 37 pins. This connector is specified for the *RS-449* interface.

DCA

Defense Communication Agency (see: *Defense Information Systems Agency*)

DCE

See: *Data Circuit-terminating Equipment, Distributed Computing Environment*

DCM

See: *Digital Circuit Multiplication*

DDN

See: *Defense Data Network*

DDN NIC

Once called "The NIC," the DDN NIC's primary responsibility was the assignment of *Internet* network addresses and *Autonomous System* numbers, the administration of the root *domain*, and providing information and support services to the Internet. Since the creation of the *InterNIC*, the DDN NIC now provides those services only for the *Defense Data Network*.

DDP

See: *Datagram Delivery Protocol*

DDP

See: *Distributed Data Processing*

DDS

See: *Dataphone Digital Service*

de facto standard

See: *standard*

deaf node

A node on a *CSMA/CD* network which does not listen (carrier sense) before transmitting. This is usually a hardware problem with the node's *interface* to the network.

DECmcc

An implementation of *Digital Network Architecture's Enterprise Management Architecture.*

DECnet

A proprietary network protocol designed by Digital Equipment Corporation. The functionality of each Phase of the implementation, such as Phase IV and Phase V, is different. Phase V is *OSI* compliant. See also: *Digital Network Architecture.*

decode

The process of transforming an *encoded* signal into the original voice or data stream. See also: *coder/decoder, modulator/demodulator.*

decoder

See: *coder/decoder*

decrypt

The process of transforming *encrypted* information into the original, readable form. See also: *Data Encryption Standard.*

DECUS

Digital Equipment Corporation User Society

DECwindows

Digital Equipment Corporation's implementation of *X windows* for Digital workstations. It includes a set of utilities to supplement the basic X windows services.

default route

A *routing* table entry which is used to direct packets addressed to networks not explicitly listed in the routing table.

default zone

The *zone* to which any node on an *AppleTalk* network will belong until it is overridden by the explicit selection of another zone.

Defense Advanced Research Projects Agency (DARPA)

An agency of the United States Department of Defense responsible for the development of new technology for use by the military. DARPA (formerly known as ARPA) was responsible for funding much of the development of the *Internet* as it is known today, including the *Berkeley* version of *UNIX* and *TCP/IP.*

Defense Data Network (DDN)

A global communications network serving the United States Department of Defense composed of *MILNET,* other portions of the *Internet,* and classified networks which are not part of the Internet. The DDN is used to connect military installations and is managed by the *Defense Information Systems Agency.*

Defense Information Systems Agency (DISA)

Formerly called the Defense Communications Agency (DCA), this is the government agency responsible for managing the *Defense Date Network* portion of the *Internet,* including the *MILNET.* Currently, DISA administers the *DDN,* and supports the user assistance services of the *DDN NIC.*

DEK

See: *Data Encryption Key*

DELNI

See: *Digital Local Network Interconnect*

demodulator

See: *modulator/demodulator*

demultiplexor

See: *multiplexor*

DES

See: *Data Encryption Standard*

designated router

A *DECnet router* which is the default router for all *End Stations* on a *Local Area Network.* That is, an End Station which needs to send a packet to a node not on the *LAN* sends it to the designated router.

device driver

The *datalink-layer* software required to operate a device (e.g., network interface card). Device drivers are specific for the hardware for which they are written, but may share a common high-level interface (i.e., the *network-layer* interface).

DNCP

See: *Dynamic Host Configuration Protocol*

DIA/DCA

See: *Document Interchange Architecture/Document Content Architecture*

dial backup

A feature of some network nodes which allows a *dial-up* connection to be initiated in the event of a failure of the primary, typically *leased line*, connection.

dial-up

A temporary, as opposed to dedicated, connection between machines established over a standard phone line. See also: *leased line.*

dialog management

The control of which node may transmit on a *Half Duplex* connection. *Full Duplex* connections do not require dialog management because data may be transmitted in both directions simultaneously.

digital

A discrete form of transmission signal based on transmission of binary data. See also: *analog, electrical signaling.*

Digital Access and Cross-connect System (DACS)

Central Office switching equipment which allows *T1* channels or subchannels to be switched or cross-connected to other T1 channels or subchannels.

Digital Circuit Multiplication (DCM)

A method of increasing the effective *carrying capacity* of *P-ISDN* communication channels using speech encoding at 64Kb/s.

Digital Local Network Interconnect (DELNI)

Digital's 8-port *Ethernet concentrator.* Often called "Ethernet in a Can" it was one of the first Ethernet concentrators. It may be operated standalone, as an Ethernet with only eight *transceivers,* or as a concentrator replacing eight transceiver *taps.*

Digital Network Architecture (DNA)

Digital Equipment Corporation's networking *architecture.* The *DECnet* protocol suite is a DNA implementation.

Digital Termination Systems (DTS)

A microwave-based transmission technology for line-of-sight applications where it is too difficult or costly to run cable.

digitize

To convert *analog* data or signaling into *digital* data or signaling. See also: *coder/decoder, Optical Character Recognition.*

Direct Memory Access (DMA)

The ability of devices to access memory without *CPU* intervention.

directed broadcast

A *broadcast* packet sent by a network node to all nodes on all directly attached networks (usually *LAN*s). All nodes on those networks receive the packet, but do not *forward* it any further. See also: *network broadcast, subnet broadcast. multicast.*

directory

A catalog of files, people, or *services.* See also: *Hierarchical File System, white pages, whois, X.500.*

Directory Access Protocol (DAP)

X.500 protocol used for communications between a *Directory User Agent* and a *Directory System Agent.*

Directory System Agent (DSA)

The software which provides the *X.500* directory service for a portion of the directory information base. Generally, each DSA is responsible for the directory information for a single organization or organizational unit.

Directory User Agent (DUA)

The software that accesses the *X.500* directory service on behalf of the directory user. The directory user may be a person or another software element.

DIS

See: *Draft International Standard*

DISA

See: *Defense Information Systems Agency*

Disk Operating System (DOS)

An *Operating System* which operates on *Personal Computers.* The two most widely deployed DOS implementations are *MS-DOS* and *PC-DOS.*

diskless node

A network node which has no local disk storage attached. It depends on disk servers and *file servers* for storage space. See also: *Network File System, Remote File System.*

Distance-Vector algorithm (DV algorithm)

A *route* information distribution algorithm used by many *routing protocols.* It operates by locally distributing global route information. That is, each router knows how to reach every *subnet* in the network (or network in the internet), and it shares that information with each of its *neighbors.* See also: *IGRP, RIP, RTMP, Link-State algorithm.*

Distributed Computing Environment (DCE)

An *architecture* of standard programming *interfaces,* conventions, and server functions (e.g., naming, *Distributed File System, Remote Procedure Call*) for distributing applications transparently across networks of heterogeneous computers. DCE is promoted and controlled by the *Open Software Foundation,* a consortium led by Digital, IBM and Hewlett Packard.

Distributed Data Processing (DDP)

A collection of *file servers, print servers,* data processors, and *hosts* which are networked together in such a way as to allow applications running on the hosts to operate as though the resources were local.

distributed database

A collection of several different data repositories which appears as a single database to the user. In the *Internet,* the *Domain Name System* and *Gopher* are examples of distributed databases.

DIX Ethernet

See: *Ethernet*

DMA

See: *Direct Memory Access*

DNA

See: *Digital Network Architecture*

DNIC

See: *Data Network Identification Code*

DNS

See: *Domain Name System*

Document Interchange Architecture/Document Content Architecture (DIA/DCA)

IBM's *SNA architectures* for the transmission and storage of data, voice and video.

DoD

United States Department of Defense

domain

A generic term for a logical collection of entities. For example, a *routing* domain is a collection of *routers* which communicate using a common *IGP.* See also: *Administrative Domain, Domain Name System.*

Domain Name System (DNS)

The DNS is a general purpose distributed, replicated, data query service. Its principal function is the resolution of *Internet addresses* based on *Fully Qualified Domain Names.* Some important domains are: .COM (commercial), .EDU (educational), .NET (network operations), .GOV (United States government), and .MIL (United States military). All countries also have a domain. For example, .US (United States), .UK (United Kingdom), .AU (Australia). See also: *Fully Qualified Domain Name, name resolution.*

DOS

See: *Disk Operating System*

dot address (dotted decimal notation)

A reference to the common notation for *IP addresses* of the form A.B.C.D; where each letter represents, in decimal, one byte of a four byte IP address.

down-link

The signal sent from a satellite to an *earth station.* It is usually a transformation of an *up-link* signal. See also: *transponder.*

downstream neighbor

A term indicating the *neighbor* from which a node receives a packet in a *ring network.* Every node is the downstream stream neighbor of its *upstream neighbor.*

downtime

An reference to the amount of time a computer has been inoperable, or a network unavailable. See also: *uptime.*

Draft International Standard (DIS)

An *ISO* document which is one step away from becoming an *International Standard.* Typically, a DIS is technically sound and only needs minor modifications before becoming an *IS.*

driver

See: *device driver*

drop cable

The cable which connects a network communications device to a *Local Area Network.* A drop cable should not be confused with a *transceiver cable.*

DS-0

A 64Kb/s *digital* communications *channel.* A DS-0 is typically one *sub-channel* in a *T1* or *T3* communications channel.

DS-1

A framing specification for *T1 synchronous* communication lines. See also: *Extended Superframe Format.*

DS-3

A framing specification for *T3 synchronous* communication lines. See also: *Extended Superframe Format.*

DSA

See: *Directory System Agent*

DSAP

Destination Service Access Point (see: *Service Access Point*)

DSR

See: *Data Set Ready*

DTE

See: *Data Terminal Equipment*

DTMF

See: *Dual Tone Multi-Frequency*

DTR

See: *Data Terminal Ready*

DTS

See: *Digital Termination Systems*

du jure standard

See: *standard*

DUA

See: *Directory User Agent*

Dual Tone Multi-Frequency (DTMF)

The signaling method used by Touch-Tone telephones. Each key generates two tones, the combination of which is unique for each of the 12 keys.

DV-algorithm

See: *Distance-Vector algorithm*

Dynamic Adaptive Routing (DAR)

Automatic rerouting of traffic based on a sensing and analysis of current, actual network conditions. DAR does not include *routing* decisions based on predefined information. See also: *static route.*

dynamic bandwidth allocation

A feature of some network nodes which allows additional transmission circuits to be temporarily added to the basic allocation during a high burst of traffic. See also: *bandwidth.*

Dynamic Host Configuration Protocol (DHCP)

A *UDP*-based protocol which allows a network node to completely configure itself with out any direct user intervention. Some prior configuration may need to be done by a network administrator. DHCP is the successor to *BOOTP.*

Dynamically Assigned Socket (DAS)

A *socket* number which is dynamically assigned by *DDP* on request from a client application program. DAS numbers range from 128 to 254.

E

EARN

See: *European Academic and Research Network*

earth station

A ground-based communications device which maintains a link with an orbiting satellite. It is sometimes called a ground station. See also: *satellite communication, Very Small Aperture Terminal.*

EBCDIC

See: *Extended Binary Coded Decimal Interchange Code*

Ebone

A pan-European *backbone* network.

echo

The reflection of a transmitted signal, from the point of reception, back to the sender. See also: *terminated line.*

echo suppression

Equipment used by telephone service providers which eliminates echoing of transmitted signals. It is this equipment which is deactivated by the initial tone generated by *modems.*

ECMA

See: *European Computer Manufacturers Association*

EDI

See: *Electronic Data Interchange*

EEPROM

Electrically-Erasable Programmable Read-Only Memory

EFF

See: *Electronic Frontier Foundation*

EFT

Electronic Funds Transfer

EGP

See: *Exterior Gateway Protocol*

EHF

See: *Extremely High Frequency*

EIA

See: *Electronic Industries Association*

EID

See: *Endpoint Identifier*

ELAP

See: *EtherTalk Link Access Protocol*

electrical signaling

Any of a number of techniques used to *encode* binary data with electrical impulses. The samples shown below are described in this glossary.

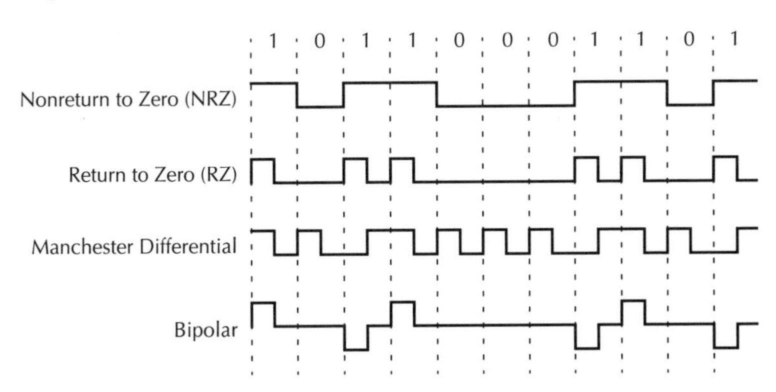

Electromagnetic Interference (EMI)

The electromagnetic radiation which leaks from a device which has been linked to a transmission *medium.* It primarily results from using high-frequency energy and signal *modulation.* See also: *Radio Frequency Interference.*

electromagnetic spectrum

The range of electromagnetic frequencies from power and telephone, to radio waves, to microwaves, to infrared, to visible light, to ultraviolet, to X-rays, to gamma rays, to cosmic rays. The frequencies from ULF to EHF are described in this glossary.

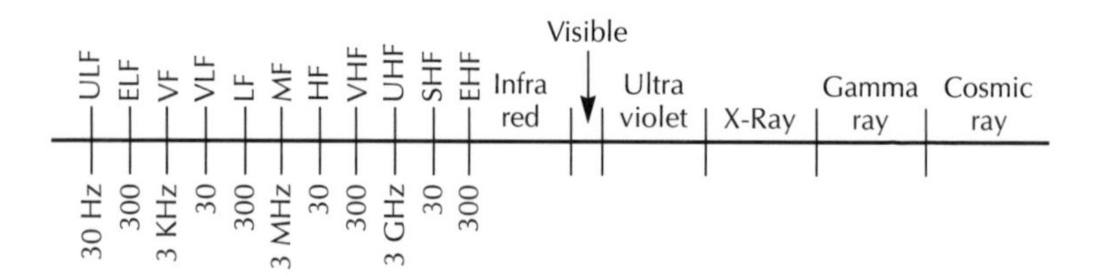

Electronic Data Interchange (EDI)

The *ANSI* standard for exchanging business documents over a data network.

Electronic Frontier Foundation (EFF)

A foundation established to address the social and legal issues arising from the impact on society of the increasingly pervasive use of computers as a means of communication and information distribution.

Electronic Industries Association (EIA)

An electronics standards making organization and a member of *ANSI. RS-232, RS-422* and *RS-449* are examples of EIA standards.

Electronic Mail (email)

A system whereby a computer user can exchange messages with other computer users (or groups of users) via a communications network. Electronic mail is one of the most popular uses of the *Internet.* See also: *mailing list, Usenet.*

Electronic Switching System (ESS)

One of AT&T's *Central Office* switching devices which uses a stored program control.

ELF

See: *Extremely Low Frequency*

EMA

See: *Enterprise Management Architecture*

email

See: *Electronic mail*

email address

The domain-based or *UUCP* address that is used to send *Electronic Mail* to a specified destination. For example, *gmalkin@xylogics.com* is the *email* address of this book's author. See also: *bang path, mail path, Domain Name System, Fully Qualified Domain Name.*

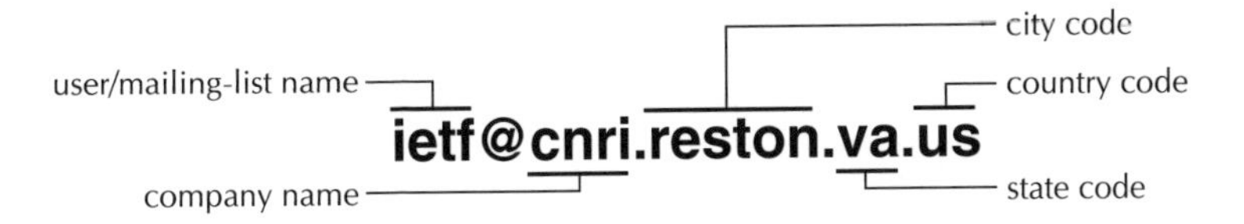

EMI

See: *Electromagnetic Interference*

emulator

A device or program which imitates all or part of another device or program. An emulator provides the same *interfaces* and responses as that which it emulates, but does not necessarily perform the same functions.

encapsulation

The technique used by *layered* protocols in which a layer adds *header* information to the *Protocol Data Unit* from the layer above. As an example, in *Internet* terminology, a packet would contain a header from the *physical layer*, followed by a header from the *network layer* (*IP*), followed by a header from the *transport layer* (*TCP*), followed by the application protocol data.

encode

The fully reversible process of transforming a voice or data stream into a format suitable transmission. See also: *decode.*

encrypt

The process of transforming readable information into an unreadable form according to a known algorithm. The process may be reversible, so that the original information may be *decrypted*; or one-way, wherein the original information cannot be recovered. See also: *encryption.*

encryption

The manipulation of a packet's data in order to prevent anyone but the intended recipient from reading that data. There are many types of data encryption, and they are the basis of network security. See also: *Data Encryption Standard.*

end node

A *DECnet* non-*routing* network *node.* See also: *End System, host.*

End Of Address (EOA)

A *control character* which marks the end of an address in a *message's header.*

End Of Block (EOB)

A *control character* which marks the end of a block of text in *character-oriented* transmissions.

End Of Message (EOM)

A *control character* which marks the end of a *message* in *character-oriented* transmissions. See also: *Start Of Message.*

End of Text (ETX)

A *control character* which marks the end of a *message's* text in *character-oriented* transmissions. See also: *Start of Text.*

End Of Transmission (EOT)

A *control character* which indicates that all user data has been transmitted.

End System (ES)

The *OSI* term for a network node which is capable of communicating through all seven layers of the *OSI reference model.* See also: *host.*

End System to Intermediate System (ES-IS)

An *OSI* protocol which *End Systems* use to identify themselves to *Intermediate Systems.*

end-to-end

A term describing the entire *path* between two communicating applications. The path may or may not include *Intermediate Systems.* See also: *hop-to-hop.*

Endpoint Identifier (EID)

A unique identification for a *node* on an *internet.* The term was coined because *address* was heavily overused and carried a lot of historical connotations.

Enterprise Management Architecture (EMA)

Digital's *network management architecture.* It defines user *interfaces* which operate on multiple displays and with multiple *protocols. DECmcc* is an implementation of EMA.

entity

Any independently addressable *node, service* or *process* which is accessible on a network. Some entities are relatively permanent (e.g., *host, router*), while others are ephemeral (e.g., one end of a *VTP* connection).

EOA

See: *End Of Address*

EOB

See: *End Of Block*

EOF

End Of File

EOM

See: *End Of Message*

EOT

See: *End Of Transmission*

ephemeral port

A *TCP* or *UDP port* number which is assigned to, usually the local end of, a connection. It is chosen so as to create a unique *socket* for one end of a connection. Ephemeral port numbers range from 1025 to 65535. See also: *well-known port.*

EPROM

Erasable Programmable Read-Only Memory

Erlang

A unit of telecommunications traffic measurement equal to 36*ccs*. It represents the maximum *carrying capacity* of a conventional communications circuit.

error correction code

A method of *encoding* data so that all or part of a transmitted or stored *message* can be reconstructed if bit errors occur. See also: *Forward Error Correction.*

Error Detection and Recovery Class

See: *Transport Class 4*

ES

See: *End System*

ES-IS

See: *End System to Intermediate System*

ESF

See: *Extended Superframe Format*

ESS

See: *Electronic Switching System*

ETB

End of Transmitted Block (see: *End Of Block*)

ethertype

The 2-octet field in the *header* of an *Ethernet* frame which indicates the *network-layer* protocol being carried in the frame.

ETX

See: *End of Text*

Ethernet

A 10Mb/s standard for *LAN*s, initially developed by Xerox, and later refined by Digital Equipment Corporation, Intel and Xerox (DIX). All hosts are connected to a *coaxial cable* where they contend for network access using *CSMA/CD*. See also: *802.3, token ring.*

Ethernet meltdown

An event that causes saturation, or near saturation, on an *Ethernet.* It usually results from illegal or mis-routed packets and typically lasts only a short time.

EtherTalk Link Access Protocol (ELAP)

The *AppleTalk, datalink-layer* protocol used over *802.3* networks. See also: *AppleTalk Remote Access Protocol, LocalTalk Link Access Protocol, TokenTalk Link Access Protocol.*

European Academic and Research Network (EARN)

The first general purpose computer network dedicated to universities and research institutions throughout Europe, the Middle East, and Africa. EARN is part of the *Internet.*

European Computer Manufacturers Association (ECMA)

An association of computer suppliers which sell equipment in Europe. ECMA is a non-voting member of *ITU-TSS.*

European Workshop for Open Systems (EWOS)

Europe's *OSI* implementors workshop. It is a peer to the *Asia and Oceania Workshop* and the *Workshop for Implementors of OSI.*

even parity

See: *parity check*

EWOS

See: *European Workshop for Open Systems*

exchange

See: *Central Office*

Exactly-Once transaction (XO transaction)

An *ATP* transaction in which the *request packet* is transmitted only once; there is no *retransmission.* This protects applications which may produce unpredictable results if a request were to be received twice.

Exchange Identification (XID)

An *HDLC* control *frame* which a *node* transmits upon attaching to a *medium.* It announces the existence of the node and contains, at least, the node's identification.

Extended Binary Coded Decimal Interchange Code (EBCDIC)
A standard character-to-number encoding used primarily by IBM computer and network systems. See also: *ASCII.*

Extended Superframe Format (ESF)
T1 24-bit frame format which is an extension to the older *D4 framing.* It provides for a *CRC,* frame synchronization, and datalink bits.

Exterior Gateway Protocol (EGP)
A protocol which distributes *routing* information to the *routers* which connect *Autonomous Systems.* The term *gateway* is historical; *router* is currently the preferred term. There is also a *routing protocol* called EGP defined in RFC 904. See also: *Autonomous System, Border Gateway Protocol, Interior Gateway Protocol.*

External Data Representation (XDR)
A standard for machine independent data structures developed by Sun Microsystems. It is similar to *Abstract Syntax Notation One.*

Extremely High Frequency (EHF)
A portion of the *electromagnetic spectrum* (microwave) with frequencies ranging from 30GHz to 300GHz.

Extremely Low Frequency (ELF)
A portion of the *electromagnetic spectrum* with frequencies ranging from 30Hz to 300Hz. These frequencies are used to transmit information to deep-ocean submarines.

F

facsimile (fax)

A text and image transmission service which operates by *digitizing* plain-paper input, transmitting the digital information across a standard telephone line, and reproducing the original input on plain-paper output. There are also several PC device cards which can read files from disk and send the files in fax format, and accept fax transmissions and write the files to disk.

FARNET

A international, non-profit organization, established in 1986 and incorporated in 1990, whose mission is to promote internetworking to support education, research, library access, health care, economic development, and citizen access. Membership is open to any organization which supports FARNET's goals.

Fast Packet

See: *Asynchronous Transfer Mode*

FAQ

Frequently Asked Question

fast select

An *X.25* option which allows data to be transmitted in the call setup and call clear packets.

fault tolerant

A description of computer and networking hardware and software products which are resistant to failures. For example, a *star network* with redundant *wiring hubs*, or a device with *hot-swap*pable modules. See also: *Mean Time Between Failures, Mean Time To Repair.*

fax
See: *facsimile*

FCC
See: *Federal Communications Commission*

FCS
See: *Frame Check Sequence*

FDDI
See: *Fiber Distributed Data Interface*

FDM
See: *Frequency Domain Multiplexing*

FDX
See: *Full Duplex*

FEC
See: *Forward Error Correction*

Federal Communications Commission (FCC)
A board of commissioners created by the Communications Act of 1934 with the authority to regulate all United States interstate communications. It is also responsible for allocating *bandwidth* in the *electromagnetic spectrum.*

Federal Information Exchange (FIX)
One of the connection points between the United States government *internets* and the *Internet.*

Federal Information Processing Standards (FIPS)
NIST specifications for computing and telecommunications equipment sold to the United States government.

Federal Networking Council (FNC)
The coordinating group of representatives from those federal agencies involved in the development and use of federal networking, especially those networks using *TCP/IP* and the *Internet.* Current members include representatives from DOD, DOE, *DARPA*, *NSF*, NASA, and HHS.

Fiber Distributed Data Interface (FDDI)
A high-speed (100Mb/s) *LAN* standard. The underlying *medium* is *fiber optics*, and the topology is a dual-attached, counter-rotating *token ring*.

fiber optics
A transmission technology which user *laser* or monochromatic *LED* light to carry information on glass or plastic fibers. See also: *multi-mode fiber, single-mode fiber.*

file server
A dedicated network node which provides mass data storage for other, perhaps *diskless*, nodes on the network.

file transfer
The copying of a file from one computer to another over a computer network. See also: *AFP, FTP, FTAM, TFTP, Kermit, xmodem, ymodem.*

File Transfer, Access, and Management (FTAM)
An *OSI* protocol which allows a user on one *End System* to access, and transfer files to and from, another End System over an OSI network.

File Transfer Protocol (FTP)
A protocol which allows a user on one *host* to access, and transfer files to and from, another host over a *TCP/IP* network. Also, FTP is usually the name of the program the user invokes to execute the protocol. See also: *anonymous FTP.*

filtering
The special handling of packets in a data stream for administrative reasons. Special handling may include the generation *audit trails*, or the discarding of packets. Filters are typically created for security reasons. See also: *firewall.*

finger
A program which displays information about a particular user, or all users, logged on the local system or on a remote system. It typically shows full name, last login time, idle time, terminal line, and terminal location (where applicable). It may also display plan or project files created by the user.

FIPS
See: *Federal Information Processing Standards*

firewall

A *gateway* between *networks* which is used to restrict the flow of information between the networks. For example, since *NFS* and *TFTP* allow a remote user to alter files on disk but do not have any *authentication* built into the protocols, as *FTP* does, a firewall may be used to exclude NFS and TFTP packets. See also: *filtering, chokepoint.*

First In First Out (FIFO)

The shorthand description of the operation of a *queue.* It indicates that the first item to be added is the first item to be removed. See also: *Last In First Out.*

FIX

See: *Federal Information Exchange*

flag

A *control character* used in many protocols to mark the beginning and end of a *message.* Many protocols (e.g., *HDLC, SDLC*) use the following 8-bit pattern as a flag: 01111110. See also: *bit stuffing.*

flame

A strong opinion and/or criticism of something, usually as a frank inflammatory statement, in an *Electronic Mail* message. It is common to precede a flame with an indication of pending fire (i.e., FLAME ON!). Flame Wars occur when people start flaming other people for flaming when they shouldn't have.

FLEA

See: *Four Letter Extended Acronym*

flow control

A mechanism with which the rate of data transmission can be regulated so as not to *overrun* the receiver. See also: *XON/XOFF, Clear To Send, credit, sliding window, source quench, congestion control.*

flush

The forcing of *queued* data out to a storage device or onto a communications medium. This is typically done to synchronize nodes, or to prepare for a graceful shutdown.

FM

See: *Frequency Modulation*

FNC

See: *Federal Networking Council*

For Your Information (FYI)

A subseries of *RFCs* which are not technical standards or descriptions of protocols. FYIs convey general information about topics related to *TCP/IP* or the *Internet.* See also: *STD.*

format

The specification of a data structure for stored or transmitted information. For example, networking *protocols* specify the format of the *headers* they use so that *peers* can read the information correctly.

Fortran

Formula Translator (early computer programming language)

forwarding

Accepting a packet, from the originator of the packet or from an intermediate *hop,* and then sending the packet to its destination or another intermediate hop. Forwarding is the action of passing along a packet and should not be confused with *routing,* which is the determination of a packet's *path* through a network. See also: *bridge, router, Intermediate System.*

Forward Error Correction (FEC)

A technique by which the receiver of a *message* may correct bit errors which occurred during transmission, without the need to ask for a retransmission. The technique requires that additional bits be transmitted, thus incurring *overhead.* See also: *Error Correction Code.*

Four Letter Extended Acronym (FLEA)

A recognition of the fact that there are far too many acronyms in the computer industry. See also: *Three Letter Acronyms.*

FQDN

See: *Fully Qualified Domain Name*

fragment

A piece of a *datagram.* When a *router* is *forwarding* an *IP* packet to a network which has an *MTU* smaller than the size of the packet being forwarded, the router is forced to break up that packet into multiple fragments. These fragments will be reassembled by IP at the destination host.

fragmentation

The *IP* process in which a packet is broken into smaller pieces to fit the requirements of a physical network over which the packet must pass. See also: *reassembly.*

frame

A *datalink-layer* packet which contains the *header* and *trailer* information required by the physical *medium.* That is, *network-layer* packets are encapsulated to become frames. See also: *datagram, encapsulation.*

Frame Check Sequence (FCS)

A 16-bit field which contains error checking information. It is usually appended to the end of a *frame.*

freenet

Community-based *Bulletin Board System* with *Electronic Mail,* information services, interactive communications, and conferencing. Freenets are funded and operated by individuals and volunteers, much like public television. They are part of the *National Public Telecomputing Network.*

Frequency Division Multiplexing (FDM)

A technique for sharing a single communications *channel* among multiple users. Each user transmits and receives data on separate frequencies, which are separated from each other by *guard frequencies.* See also: *Time Division Multiplexing.*

Frequency Modulation (FM)

A transmission method in which data is *encoded* over a *carrier* signal by variations in the frequency of the carrier. See also: *Amplitude Modulation, Phase Modulation, Pulse Width Modulation.*

FRICC

Federal Research Internet Coordinating Committee (see: *Federal Networking Council*)

front-end processor

A dedicated device which performs data communication functions for one or more hosts, thus serving to reduce the *overhead* load on the host(s).

FTAM
See: *File Transfer, Access, and Management*

FTP
See: *File Transfer Protocol*

Full Duplex (FDX)
The operation of a communications channel which allows communication devices to transmit and receive simultaneously. See also: *Half Duplex, simplex.*

fully connected network
A network *topology* in which every node is directly connected to every other node.

Fully Qualified Domain Name (FQDN)
The full, globally unique name of a system, rather than just its *hostname.* For example, *venera* is a hostname, and *venera.isi.edu* is an FQDN. See also: *Domain Name System.*

FYI
See: *For Your Information*

G

G

See: *Giga-*

gain

Increased signal power resulting from signal amplification. See also: *attenuation.*

GAP

See: *Gateway Access Protocol*

gated

Gatedaemon is a program which supports multiple *routing protocols* and protocol families. It may be used for routing, and makes an effective platform for routing protocol research. The software is freely available by *anonymous FTP* from *gated.cornell.edu.* It is often pronounced "gate-dee." See also: *EGP, IGP, BGP, RIP, IS-IS, OSPF, routed.*

gateway

The term *router* is now used in place of the original definition of *gateway.* Currently, a gateway is a communications device/program which passes data between networks having similar functions but dissimilar implementations. This should not be confused with a *protocol converter.* By this definition, a *router* is a layer 3 (*network layer*) gateway, and a mail gateway is a layer 7 (*application layer*) gateway.

Gateway Access Protocol (GAP)

A *DECnet protocol* used by an *application process* to access gateways to other protocols (e.g., *IP, SNA*).

Gb/s

Gigabits per second

geosynchronous orbit

An orbit above the equator where the altitude and speed of a satellite combine to maintain the satellite's position over the same point on the earth's surface. The altitude is approximately 23,242 statute miles, which causes a signal *propagation delay* of approximately 0.125 seconds.

gentlebeing

A humorous term used in salutations (i.e., Dear Gentlebeings) which does not discriminate against sex or species. See also: *Internaut.*

GHz

Gigahertz (one billion cycles per second)

GIF

See: *Graphical Interchange Format*

Giga- (G)

In communications (i.e., Gb/s) and computer memory, 1G equals 2^{30} (1024^3 = 1,073,741,824). In other applications, 1G equals 10^9 (1,000,000,000).

GMT

See: *Greenwich Mean Time*

go-back-N

A form of *Automatic Repeat Request.* The receiver indicates that the Nth packet has been lost or damaged. The transmitter then *retransmits* that packet and all subsequent packets. This form of *ARQ* represents a compromise between the inefficiency of *stop-and-wait* and the complexity and buffering requirements of *selective-repeat.*

Gopher

A distributed information service that makes available hierarchical collections of information across the *Internet.* Gopher uses a simple protocol that allows a single Gopher *client* to access information from any accessible Gopher *server,* providing the user with a single "Gopher space" of information. Public domain versions of the client and server are available. See also: *archie, archive site, Prospero, Wide Area Information Servers.*

GOSIP

See: *Government OSI Profile*

Government OSI Profile (GOSIP)
A subset of *OSI* standards specific to United States government procurements, designed to maximize interoperability in areas where plain OSI standards are ambiguous or allow excessive options. See also: *Manufacturing Automation Protocol, Technical and Office Protocols.*

Graphical Interchange Format (GIF)
A standard bitmap format used to store and transmit *digitized* images (pictures).

Greenwich Mean Time (GMT)
The time of day in Greenwich, England. Since Greenwich lies on the Prime Meridian (longitude 0 degrees), it is used as a time referent by networks and organizations which span multiple time zones. GMT is also referred to as Zulu time. See also: *Coordinated Universal Time.*

ground station
See: earth station

group address
An *address* which multiple network nodes will recognize. It is typically a *hardware address* or a *network-layer* address. See also: *multicast.*

guard frequency
In *Frequency Domain Multiplexing*, a portion of the *bandwidth* of a transmission *medium* which is reserved in order to create a gap between frequencies which are being used to carry information. The gap is intended to reduce *interference* between the signals.

guest
A special userid supported by many *UNIX*-like operating systems. Anyone may use a guest userid without a password. System access for guest userids is usually restricted. See also: *anonymous FTP.*

Guide to the Use of Standards (GUS)
A publication of the *Standards Promotion and Application Group* which contains *OSI* option subsets.

GUS
See: *Guide to the Use of Standards*

hacker

A person who delights in having an intimate understanding of the internal workings of a system, computers and computer networks in particular. The term is often misused in a pejorative context, where *cracker* would be the correct term.

Half Duplex (HDX)

The operation of a communications channel which permits data to be transmitted in only one direction at a time. A protocol is required to synchronize the communication devices to prevent *collisions.* See also: *dialog management, Full Duplex, simplex.*

half router

A network device which typically connects a *Local Area Network* to a serial communications line. Another half router is usually on the other end of that serial line. Together, the two half routers appear to the *LAN*s as a single *router* which resides between them.

handle

A term for a temporary identifier. For example, file handles are associated with currently open files; when the file is closed, the handle disappears.

handshake

An exchange of *control characters* or signals between two communication devices which establishes a communications channel. See also: *three-way handshake.*

hardware address

The address intrinsic to a device attached to a network. Typically, that address is "built-in" to the device by the manufacturer.

HASP
See: *Houston Automatic Spooling Program*

HDLC
See: *High-level Data Link Control*

HDX
See: *Half Duplex*

header
The portion of a *packet*, preceding the actual data, containing source and destination addresses, and error checking and other fields. A header is also the part of an *Electronic Mail* message that precedes the body of a message and contains, among other things, the message originator, date and time. See also: *trailer.*

Hertz (Hz)
A measure of the number of cycles per second of an *analog* signal. One Hz is one cycle per second.

heterogeneous network
A network running multiple *network-layer* protocols. See also: *homogeneous network, multi-protocol router.*

hexadecimal (hex)
The base-16 numbering system where the values for 10 through 15 are represented by *A* through *F*, respectively. It is commonly used for computer notation because 16 values can be represented by 4 *bits*, or one *nibble*. See also: *binary, octal.*

HF
See: *High Frequency*

HFS
See: *Hierarchical File System*

hierarchical architecture
An *architectural* paradigm for network *protocol stacks*. It is similar to a *layered architecture* except for the relaxation of two layering restrictions. First, a level (so-called to avoid confusion with *layer* in a layered architecture) may communicate directly with any other level, not just the levels directly above and below it in the protocol stack. For example, an appli-

cation may communicate directly with the network protocol, avoiding the intervening levels. Second, protocols within the same level may communicate as *peers* rather than traverse down the stack then loop back up.

Hierarchical File System (HFS)

A directory-oriented file system, where in a root directory has multiple subdirectories, which may have themselves have subdirectories, ad nauseam. Within any subdirectory, there may be files. The *UNIX* file system is an example of an HFS.

hierarchical routing

The complex problem of *routing* on large networks can be simplified by reducing the size of the networks. This is accomplished by breaking a network into a hierarchy of networks, where each level is responsible for its own routing. The *Internet* has, basically, three levels: the backbones, the mid-levels, and the stub networks. The backbones know how to route between the mid-levels, the mid-levels know how to route between the stubs, and each stub (being an *Autonomous System*) knows how to route internally. See also: *Exterior Gateway Protocol, Interior Gateway Protocol, stub network, transit network.*

High Frequency (HF)

A portion of the *electromagnetic spectrum* (short-wave) with frequencies ranging from 3MHz to 30MHz.

High-level Data Link Control (HDLC)

A *CCITT* standard, *bit-oriented, datalink-layer* protocol on which most other bit-oriented protocols are based.

High Performance Computing and Communications (HPCC)

High performance computing encompasses advanced computing, communications, and information technologies, including scientific workstations, supercomputer systems, high speed networks, special purpose and experimental systems, the new generation of large scale parallel systems, and application and systems software with all components well integrated and linked over a high speed network.

High Performance Parallel Interface (HIPPI)

An *ANSI* standard which extends the computer bus over fairly short distances at speeds of 800Mb/s and 1600Mb/s. HIPPI is often used in a computer room to connect a supercomputer to *routers*, frame buffers, mass-storage peripherals, and other computers.

HIPPI

See: *High Performance Parallel Interface*

homogeneous network

A network which runs only a single *network-layer* protocol. See also: *heterogeneous network.*

hop

An node through which a packet passes on the *path* between the packet's originating node and its destination node. See also: *Intermediate Station, router.*

hop-by-hop routing

See: *routing*

hop-to-hop

A term referring to the path between two *neighbors* on a *Local Area Network* or *point-to-point* network. Specifically, it indicates that there are no *Intermediate Systems* between the hops. See also: *end-to-end.*

host

The *Internet* term for a network node which is capable of communicating at the application layer. See also: *End System.*

host address

See: *internet address*

host byte order

The *byte order* used by the *CPU* on a given network node. See also: *network byte order, big endian, little endian.*

host number

See: *host address*

hostname

The human-readable name given to a machine. See also: *Fully Qualified Domain Name.*

hot-swap

The ability to replace a module in a device (e.g., *hub*) without having to turn the power to the device off. This allows other modules within the device to continue operating while a repair is being made. See also: *fault tolerant, Mean Time Between Failures, Mean Time To Repair.*

Houston Automatic Spooling Program (HASP)

An early (pre-*JCL*) IBM control protocol used for transmitting data processing files and jobs to mainframes.

HPCC

See: *High Performance Computing and Communications*

HTML

See: *Hypertext Markup Language*

hub

A communications device composed of several other devices. It is usually a chassis which houses multiple independent modules (e.g., terminal servers, *concentrators*). Each of the modules is provided with power and data interconnectivity.

hypertext

A form of electronic document in which indicated words within the text may be used to access another part of the document, or another document entirely. For example, this glossary highlights words which are themselves glossary entries. If this were a hypertext document, one of those words could be selected, usually by a pointing device (e.g., mouse), and that entry would then appear. Hypertext documents are created with *Hypertext Markup Language.*

Hypertext Markup Language (HTML)

The text processing language used to create *hypertext* documents. In addition to standard text processing abilities, HTML allows the creation of links which associate specified words with additional text.

Hz

See: *Hertz*

I

I-D

See: *Internet-Draft*

IAB

See: *Internet Architecture Board*

IANA

See: *Internet Assigned Numbers Authority*

IBM cabling system

The set of IBM specifications for the types of cable used to interconnect IBM networking products.

IC

See: *Integrated Circuit*

ICMP

See: *Internet Control Message Protocol*

idle

A term usually applied to data communications circuits which are not currently carrying user data. See also: *quiescent*.

IDRP

See: *Inter-Domain Routing Protocol*

IDPR

See: *Inter-Domain Policy Routing*

IEC

See: *Interexchange Carrier*

IEEE
See: *Institute of Electrical and Electronics Engineers*

IEEE 802
See: *802.x*

IEN
See: *Internet Experiment Note*

IESG
See: *Internet Engineering Steering Group*

IETF
See: *Internet Engineering Task Force*

IINREN
See: *Interagency Interim National Research and Education Network*

IGMP
See: *Internet Group Multicast Protocol*

IGP
See: *Interior Gateway Protocol*

IGRP
See: *Internet Gateway Routing Protocol*

IMAP
See: *Internet Message Access Protocol*

IMHO
In My Humble Opinion

IML
Initial Microcode Load

IMP
See: *Interface Message Processor*

impedance
The effect of capacitance, inductance and resistance (all properties of the transmission *medium*) on a transmitted signal.

implementors' agreement

A document which describes implementation details necessary to ensure *interoperability* between vendors' products. These agreements are necessary where the *standard* being implemented is vague, complex, or offers multiple implementation choices.

IMR

See: *Internet Monthly Report*

in-band signaling

The inclusion of control information on the same channel as the data, typically using *control characters*. See also: *out-of-band signaling*.

information superhighway

The media catch phrase for the *Internet*, and what the Internet will become over the next few years. See also: *National Educational and Research Network, National Information Infrastructure*.

Information Systems (IS)

The generic term for data processing and communications technologies. The term "Information Technology" has the same generic meaning.

infrared

A portion of the *electromagnetic spectrum* used on some *fiber-optic* media. It is also used for short distance, line-of-sight transmission.

Institute of Electrical and Electronics Engineers (IEEE)

A professional society which creates *standards* for the *physical layer* and *datalink layer* of the *OSI reference model*. IEEE is a member of the *American National Standards Institute*.

INTAP

See: *Interoperability Technology Association for Information Processing*

Integrated Circuit (IC)

An electronic component, containing dozens to thousands of transistors, which performs a specific function. A microprocessor is an IC, albeit a highly complex one.

Integrated Services Digital Network (ISDN)

A technology which is beginning to be offered by the telephone carriers of the world. ISDN combines voice and digital network services in a single medium, making it possible to offer customers digital data services as well as voice connections through a single cable. The standards that define ISDN are specified by *CCITT.* See also: *B-ISDN, P-ISDN.*

Integrated Voice/Data Terminal (IVDT)

A device which combines the features of a data terminal (e.g., a keyboard) and a telephone (e.g., a handset). An IVDT may also contain a local processor.

inter-domain router

A (border) *router* which *forwards* traffic between *Autonomous Systems* (routing *domains*). Routing information is exchanged with an *Exterior Gateway Protocol.* See also: *Border Gateway Protocol, intra-domain router.*

Inter-Domain Routing Protocol (IDRP)

An *OSI routing protocol* used to exchange routing information between *Autonomous Systems.* It is also being modified for use as an *EGP* on *IP internets.* See also: *Border Gateway Protocol.*

Inter-Domain Policy Routing (IDPR)

A set of *routing protocols* used to exchange routing information between *Autonomous Systems.* It differs from other *EGPs* in that network administrator policy is factored into the creation of *routes* along with routing *metrics.* See also: *Border Gateway Protocol.*

interactive

A *real-time* connection in which user activity is interleaved with computer and network responses. See also: *batch.*

Interagency Interim National Research and Education Network (IINREN)

An evolving operating network system. Near term (1992-1996) research and development activities will provide for the smooth evolution of this networking infrastructure into the future gigabit *NREN.*

Interexchange Carrier (IEC)

An *FCC* approved communication service provider which is authorized to carry *interLATA* transmissions. AT&T Communications, MCI, and Sprint are IECs.

interface (hardware)

A generic term for a *node's* access point to a physical *medium.* See also: *tap, transmitter/receiver.*

interface (software)

The boundary across which two devices or protocols communicate. An interface is characterized by the inputs a device or *protocol* requires and the responses it provides. See also: *Application Program Interface, Service Access Point.*

Interface Data Unit (IDU)

The combination of *Protocol Control Information* and a *Service Data Unit.* The *PCI* is used to create the *header* for the Layer-N *PDU.*

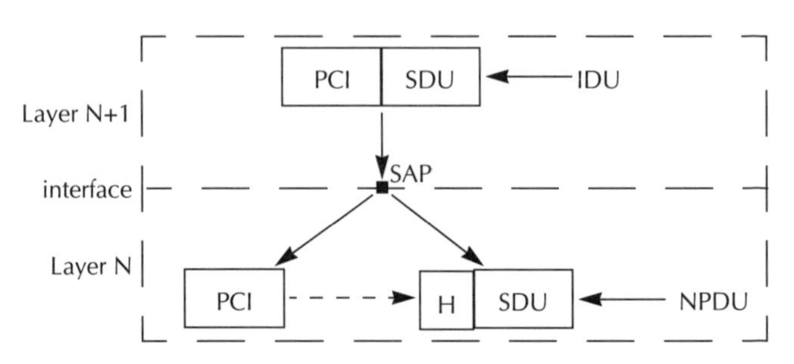

Interface Message Processor (IMP)

A device used in the *ARPANET* to interconnect various *segments* of the network. It was the precursor to modern *routers.* See also: *TIP.*

interference

The confusion of signals (electrical, radio and light) caused by unwanted interactions between signals or the occurrence of noise. See also: *Electromagnetic Interference, Radio Frequency Interference.*

Interior Gateway Protocol (IGP)

A protocol which distributes routing information to the routers within an autonomous system. The term *gateway* is historical, as *router* is currently the preferred term. See also: *Autonomous System, EGP, OSPF, RIP, RTMP.*

interLATA

See: *Local Access and Transport Area*

interleaving

The alternated, or rotated, insertion of data segments from two, or more, independent data streams into a single data stream. See also: *Time Division Multiplexing.*

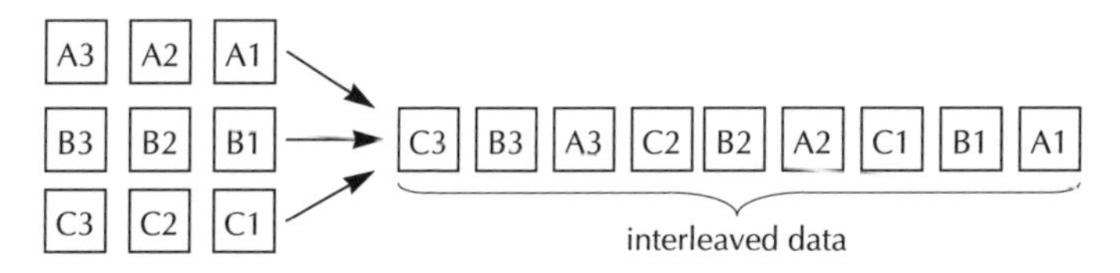

Intermediate System (IS)

The *OSI* term for a network *node* which performs *forwarding* at the *network layer.* See also: *gateway, router.*

Intermediate System to Intermediate System (IS-IS)

The *OSI* protocol with which *Intermediate Systems* exchange *routing* information. See also: *IGP, Link-State algorithm, OSPF.*

Internal Organization of the Network Layer (IONL)

The *OSI* standard for the detailed *architecture* of the *network layer.*

International Organization for Standardization (ISO)

A voluntary, non-treaty organization founded in 1946 which is responsible for creating international standards in many areas, including computers and communications. Its members are the national standards organizations of the 89 member countries, including *ANSI* for the United States. See also: *Open Systems Interconnection.*

International Standard (IS)

An *ISO* document which has been duly determined to be technically correct and complete, and therefore a *standard.* See also: *Draft International Standard.*

International Telecommunications Union (ITU)

An agency of the United Nations which coordinates the various national telecommunications standards so that people in one country can communicate with people in another country. The telephone and data communications arm of the ITU is the Telecommunications Standards Sector, which was previously known as *CCITT.*

International Telecommunications Union—Telecommunications Standards Sector (ITU-TSS)

The new name for *CCITT* since the *ITU* reorganization. The function is the same; only the name has been changed.

International Time Bureau (BIH)

The organization responsible for maintaining *Coordinated Universal Time.* See also: *Greenwich Mean Time.*

Internaut

A user of the *Internet* or member of the *Internet Society.* The term was coined in response to an *email* debate over a *netiquette* issue. The debate was over the salutation used for messages sent to *mailing lists*; "Dear Sirs" excludes women, "Dear Sirs/Madams" is too awkward, and "To Whom it may concern" is too impersonal. Many email messages are now addressed to "Dear Internauts."

internet

While an internet is a *network*, the term is usually used to refer to a collection of networks interconnected with *routers.* An internet is composed of networks which are composed of *subnets.*

Internet (note the capital "I")

The largest *internet* in the world. It is a three level hierarchy composed of *backbone* networks (e.g., *NSFNET*), *mid-level* networks (e.g., NEARnet), and *stub* networks. The Internet is a multi-protocol internet. See also: *transit network.*

internet address

An *IP address* which uniquely identifies a node on an *internet.* An Internet address (capital "I"), uniquely identifies a node on the *Internet.*

Internet Architecture Board (IAB)

The technical body which oversees the development of the *Internet* suite of protocols. It is composed of two task forces: the *Internet Engineering Task Force* and the *Internet Research Task Force.* In June of 1992, the IAB was chartered as a component of the *Internet Society* with the following responsibilities: selection of *IESG* members and chair, Internet architectural oversight, standards process oversight and appeal, *RFC* editorial management and *IANA* administration, liaison with other standards making bodies, and the providing of advice and technical guidance to the *ISOC* Board of Trustees.

Internet Assigned Numbers Authority (IANA)

The central registry for various *Internet Protocol* parameters, such as *port*, protocol and enterprise numbers, and options, codes and types. The currently assigned values are listed in the *Assigned Numbers* RFC (STD 1). To request a number assignment, contact the IANA at *iana@isi.edu*. See also: *assigned numbers, STD, Internet Registry.*

Internet Control Message Protocol (ICMP)

An ancillary protocol to *IP.* It allows for the generation and recognition of error *messages*, test messages and informational messages which extend the basic IP functionality. See also: *Packet Internet Groper.*

Internet-Draft (I-D)

Internet-Drafts are working documents of the *IETF,* its Areas, and its Working Groups. As the name implies, Internet-Drafts are draft documents. They are valid for a maximum of six months and may be updated, replaced, or obsoleted by other documents at any time. Very often, I-Ds are precursors to *RFCs*.

Internet Engineering Steering Group (IESG)

A body composed of the *IETF* Area Directors and the IETF Chair. It provides the technical review of *Internet* Standards and is responsible for day-to-day "management" of the IETF. See also: *Internet Architecture Board, Internet Research Steering Group.*

Internet Engineering Task Force (IETF)

A large, open community of network designers, operators, vendors, and researchers whose purpose is to coordinate the operation, management and evolution of the *Internet*, and to resolve short-range and mid-range protocol and architectural issues. It is a major source of proposals for protocol standards which are submitted to the *IESG* for approval. The IETF meets three times a year and extensive minutes are included in the IETF Proceedings. See also: *Internet Architecture Board, Internet Research Task Force, Internet Society.*

Internet Experiment Note (IEN)

A series of reports pertinent to the *Internet.* IENs were published in parallel to *RFCs* and are no longer active. See also: *Internet-Draft.*

Internet Gateway Routing Protocol (IGRP)

A proprietary *routing protocol* used by Cisco Systems's *routers*. It is similar to *RIP*, but carries additional information to make more complex routing decisions. See also: *IGP, IS-IS, OSPF.*

Internet Group Multicast Protocol (IGMP)

The *protocol* which specifies a *multicast* extension to the basic *Internet Protocol.* It is recommended for nodes which intend to support multicasting.

Internet Message Access Protocol (IMAP)

An *application-layer* protocol which allows a user to access and manipulate *Electronic Mail* on a remote server. See also: *RFC 822, Multipurpose Internet Mail Extensions*

Internet Monthly Report (IMR)

Published monthly, the purpose of the Internet Monthly Reports is to communicate to the *Internet* research and development groups the accomplishments and milestones reached, or problems discovered, by the participating organizations. See also: *Internet Engineering Task Force, Internet Research Task Force, Internet Society.*

internet number

See: *internet address*

Internet Protocol (IP)

The *network-layer* protocol for the *TCP/IP Protocol Suite*, defined in RFC 791. It is a *connectionless*, best-effort packet switching protocol.

Internet Protocol Control Protocol (IPCP)

The control *protocol* used to configure, enable and disable operation of *IP* over the *Point-to-Point Protocol.* Other *network-layer* protocols may have a similar *PPP* control protocol.

Internet Registry (IR)

The organization responsible for assigning identifiers, such as IP *network numbers* and *Autonomous System* numbers, to networks. The IR also gathers and registers such assigned information. At its discretion, the IR may delegate assignment responsibility for portions of the assignment space, as it has done for the *APNIC* and *RIPE NCC.* At present, the Registration

Services portion of the *InterNIC*, at Network Solutions, Inc., serves as the IR. See also: *Internet Assigned Numbers Authority.*

Internet Relay Chat (IRC)

A world-wide "party line" *protocol* which allows one to converse with others in real time. IRC is structured as a network of *servers*, each of which accepts connections from *client* programs, one per user. See also: *talk.*

Internet Research Steering Group (IRSG)

The "governing body" of the Internet Research Task Force. See also: *Internet Architecture Board, Internet Engineering Steering Group.*

Internet Research Task Force (IRTF)

An organization chartered by the *IAB* to consider long-term *Internet* issues from a theoretical point of view. It has Research Groups, similar to the *IETF*'s *Working Groups*, which are each tasked to discuss different research topics. Multi-cast audio/video conferencing and *Privacy Enhanced Mail* are examples of IRTF output.

Internet Society (ISOC)

Anon-profit, professional membership organization which facilitates and supports the technical evolution of the *Internet*, stimulates interest in and educates the scientific and academic communities, industry and the public about the technology, uses and applications of the Internet, and promotes the development of new applications for the system. The Society provides a forum for discussion and collaboration in the operation and use of the global Internet infrastructure. The Internet Society publishes a quarterly newsletter, the Internet Society News, and holds an annual conference, INET. The development of Internet technical standards takes place under the auspices of the Internet Society with substantial support from the *Corporation for National Research Initiatives* under a cooperative agreement with the United States Federal Government. See also: *IAB, IETF, IRTF.*

Internetwork Packet Exchange (IPX)

Novell's protocol used by Netware. A *router* with IPX routing can interconnect *LANs* so that Novell Netware *clients* and *servers* can communicate.

InterNIC

A five year project partially supported by the *National Science Foundation* to provide network information services to the networking community.

The InterNIC began operations in April of 1993 and is a collaborative project of three organizations: General Atomics, which provides Information Services from their location in San Diego, CA; AT&T, which provides Directory and Database Services from South Plainsfield, NJ; and Network Solutions, Inc., which provides Registration Services from their headquarters in Herndon, VA. Services are provided via the *Internet*, and by telephone, FAX, and hardcopy.

interoperability

The ability of software and hardware on multiple machines from multiple vendors to communicate meaningfully.

Interoperability Technology Association for Information Processing (INTAP)

A technical organization with the charter to develop Japanese *OSI* profiles and conformance tests.

intra-domain router

A *router* which *forwards* packets within an *Autonomous System* (routing *domain*). Routing information is exchanged with an *Interior Gateway Protocol.* See also: *IS-IS, OSPF, RIP, RTMP.*

intraLATA

See: *Local Access and Transport Area*

I/O

Input/Output

IONL

See: *Internal Organization of the Network Layer*

IP

See: *Internet Protocol*

IP address

The 32-bit address defined by the *Internet Protocol.* It is usually represented in dotted decimal notation. See also: *dot address, internet address, network address, subnet address, host address.*

IP datagram

See: *datagram*

IP:ng

A reference to IP:Next Generation. The term is used because, currently, the *protocol* to be standardized by the *IETF* as the next version of *IP* has not been determined. See also: *IPv4.*

IPCP

See: *Internet Protocol Control Protocol*

IPL

Initial Program Load

IPv4

The currently deployed version of the *Internet Protocol.* See also: *IP:ng.*

IPX

See: *Internetwork Packet Exchange*

IPX address

The combination of 32-bit network number and 48-bit host number which uniquely identifies a *node* in an *IPX* internet.

IR

See: *Internet Registry*

IRC

See: *Internet Relay Chat*

IRSG

See: *Internet Research Steering Group*

IRTF

See: *Internet Research Task Force*

IS

See: *Intermediate System, International Standard, Information Systems*

IS-IS

See: *Intermediate System to Intermediate System*

ISDN

See: *Integrated Services Digital Network*

ISO

See: *International Organization for Standardization*

ISO Development Environment (ISODE)

A software package which allows *OSI* services to use a *TCP/IP* network. ISODE is typically used by organizations developing OSI applications. It is often pronounced "eye-so-dee-eee."

ISOC

See: *Internet Society*

ISODE

See: *ISO Development Environment*

IT

Information Technology (see: Information Systems)

ITU

See: *International Telecommunications Union*

ITU-TSS

See: *International Telecommunications Union—Telecommunications Standards Sector*

IVDT

See: *Integrated Voice/Data Terminal*

J

jabber

The transmission of a continuous stream of random data on a communications channel. This is usually the result of a malfunctioning device.

jack

The receptacle into which an *RJ-11*, *RJ-12* or *RJ-45* connector is inserted.

jamming

The creation of *interference* with the intent of inhibiting communications between other communications devices. *Radio Frequency* jamming is a violation of *FCC* regulations.

JCL

See: *Job Control Language*

jitter

Slight oscillations in frequency or phase which can cause transmission errors and loss of synchronization between communication devices. Higher speed channels are more susceptible to jitter than lower speed channels.

Job Control Language (JCL)

Special instructions (control cards) prepended to an IBM application which configure the application's operating environment (e.g., open or allocate files).

jumper

A patch cable used to establish a circuit. It is usually a temporary connection created for diagnostic or testing purposes.

K

K

See: *Kilo-*

Ka band

A portion of the *electromagnetic spectrum* with frequencies ranging from 18GHz to 30GHz.

KA9Q

The amateur radio callsign of the creators of KA9Q *NOS*, a popular implementation of *TCP/IP* and associated protocols for amateur *packet radio* systems.

Kb/s

Kilobits per second

keep-alive

A process wherein one side of a connection will periodically send a packet to the remote side just to make sure the connection is still up. Such a probe is generally only sent if the line has been *idle* for some period time, because the applications have nothing to say to each other (e.g., a remote login session and the user has gone to lunch). See also: *tickle.*

Kerberos

The security system of MIT's Project Athena. It is based on symmetric key cryptography. See also: *encryption.*

Kermit

A popular *file transfer* protocol developed by Columbia University. Because Kermit operates in most operating environments, it provides an easy method of file transfer. See also: *xmodem, ymodem, zmodem, FTP, FTAM.*

key management

The creation, distribution and exchange of cryptographic keys.

Kilo- (K)

In communications (i.e., Kb/s) and computer memory, 1K equals 2^{10} (1024). In other applications 1K equals 10^3 (1000).

kiss-of-death packet

A bogus or *Martian* packet which causes the receiver to misbehave, lock up or die. Generally, a receiver should be *bullet-proofed* against such packets.

Knowbot

An experimental directory service. See also: *white pages, whois, X.500.*

Ku band

A portion of the *electromagnetic spectrum* with frequencies ranging from 10GHz to 12GHz. Ku band is used for satellite communications.

L

L band

A portion of the *electromagnetic spectrum* with a frequency of approximately 1GHz.

LAN

See: *Local Area Network*

land line

A communications *channel* which is based solely on cables. That is, it does not use any *open-air transmission* or *satellite communications.* See also: *terrestrial.*

LAP

See: *Link Access Procedure*

LAP-B

See: *Link Access Procedure-Balanced*

LAP-D

See: *Link Access Procedure-D*

laser

A device which generates a coherent, monochromatic beam of light which is used for *single-mode, fiber-optic* transmissions. LASER is an acronym for Light Amplification by Stimulated Emission of Radiation.

Last In First Out (LIFO)

The shorthand description of the operation of a *stack.* It indicates that the last item to be added is the first item to be removed. See also: *First In First Out.*

LAT

See: *Local Area Transport*

LATA

See: *Local Access and Transport Area*

late collision

A *collision* on a *CSMA/CD* transmission *medium* which occurs after a node has transmitted more than the minimum required number of bytes for a packet. It usually indicates that a *deaf node* is on the network.

latency

The time interval between a device's request for access to a resource and the time that access is granted. See also: *transmission latency*.

layer

Communication networks for computers may be organized as a set of more or less independent protocols, each in a different layer (also called level). The lowest layer governs direct host-to-host communication between the hardware at different hosts; the highest consists of user applications. Each layer builds on the layer beneath it. For each layer, programs at different hosts use protocols appropriate to the layer to communicate with each other. *TCP/IP* has five levels of protocols; *OSI* has seven. The advantages of different layers of protocols is that the methods of passing information from one layer to another are specified clearly as part of the *protocol suite*, and changes within a protocol layer are prevented from affecting the other layers. This greatly simplifies the task of designing and maintaining communication programs.

layered architecture

An *architectural* paradigm in which various protocols comprise distinct layers in a networking *protocol stack*. See also: OSI reference model, hierarchical architecture.

LCR

Least Cost Routing (see: Automatic Route Selection)

LCP

See: *Link Control Protocol*

learning bridge

A *bridge* which listens to the *broadcast medium* to which it is attached and remembers the *MAC* addresses of the nodes on that network. After a specified period of time, the bridge begins *forwarding* frames. However, it only forwards those *frames* which are addressed to nodes it did not learn during its listening period. Many learning bridges continue to learn after forwarding has begun. Some learning bridges bypass the listening phase and forward all frames until the bridge learns otherwise.

leased line

A dedicated telephone circuit for which a subscriber pays a monthly fee, regardless of utilization. It is typically a voice grade circuit which can transmit data at speeds up to 14.4Kb/s. See also: *conditioned line, dial-up.*

Least Significant Bit/Byte (LSB)

The right-most bit in a byte, or the right-most byte in a *big-endian* word. That is, the bit or byte which has the minimum value. See also: *Most Significant Bit/Byte.*

LED

See: *Light Emitting Diode*

LEN

See: *Low Entry Networking*

level 1 router

See: *intra-domain router*

level 2 router

See: *inter-domain router*

LF

See: *Low Frequency*

LIFO

See: *Last In First Out*

Light Emitting Diode (LED)

An electronic device which converts an electrical signal into a *light signal.* In addition to their use as status indicators on most computer and communications devices, LEDs are also used as the light signal source for some *multi-mode, fiber-optic* transmissions. See also: *PIN.*

light signal

A *modulated* or pulsed beam of light traveling through the air or through a *fiber-optic* cable. See also: *open-air, laser, LED.*

lightspeed

The speed at which light travels through a vacuum, 186,242.4 miles/second.

line hit

A burst of *interference* which causes data, currently being transmitted, to be corrupted. It can also cause the injection of spurious signals into a transmission *medium.*

link

A generic term for any data communications *medium* to which a network node is attached. See also: *Local Area Network, Metropolitan Area Network, Wide Area Network, point-to-point.*

Link Access Procedure (LAP)

Generically, any protocol which is manages the transmission of data across a physical *medium.* LAP is also a *datalink-layer* protocol specified for use in *X.25.* See also: *datalink layer, Logical Link Control, Media Access Control.*

Link Access Procedure-Balanced (LAP-B)

The *X.25, datalink-layer* protocol. It is almost identical to *HDLC.*

Link Access Procedure-D channel (LAP-D)

The *data-link layer* protocol used on the *D channel* of *ISDN* circuits.

Link Control Protocol (LCP)

A protocol used by *PPP* to establish, configure and test the datalink connection. It is during the LCP phase that security options (e.g., *PAP, CHAP*) are negotiated.

Link-State algorithm (LS-algorithm)

A *routing* information distribution algorithm used by many routing protocols. It operates by globally distributing local route information. That is, each *router* advertises to every other router in the network (or internet), the state of each of the *links* to which it is directly connected. See also: *IS-IS, OSPF, Distance-Vector algorithm.*

listserv

An automated *mailing list* distribution system originally designed for the *Bitnet/EARN* network. It allows users to join and leave mailing lists, and it provides *file server* functions and database searching capabilities. It is in wide use in many networks today. See also: *Usenet, Electronic Mail.*

little-endian

A format for storage or transmission of binary data in which the *Least Significant Bit/Byte* comes first. See also: *byte order, big-endian.*

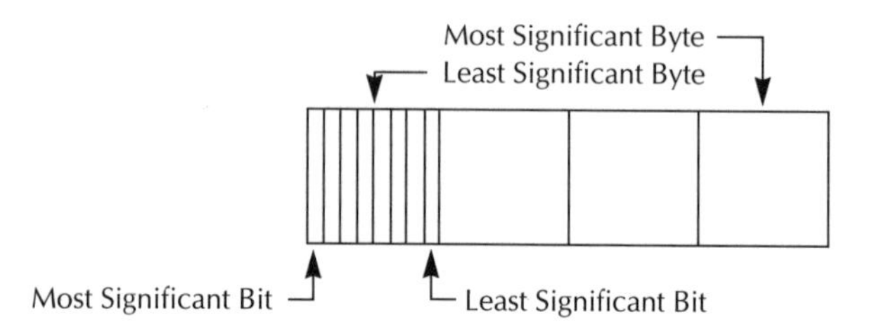

LLAP

See: *LocalTalk Link Access Protocol*

LLC

See: *Logical Link Control*

lobe

The loop of cable which connects a network node to a *wiring hub* on a *star network* or *star-wired network.*

Local Access and Transport Area (LATA)

One of 161 telephone service areas within the United States. Communication between subscribers within a single LATA is referred to as intraLATA communications. Communication between subscribers in different LATAs, adjacent or not) is referred to as interLATA.

Local Area Network (LAN)

A data *network* intended to serve an area of only a few square kilometers or less. Because the network is known to cover only a small area, optimizations can be made in the network signaling protocols which permit data rates up to 100Mb/s. See also: *802.3, Ethernet, Fiber Distributed Data Interface, token bus, token ring, Wide Area Network.*

Local Area Transport (LAT)

A *DECnet* protocol which allows users on a local host to access *services* (e.g., remote login) on a remote host. The *interface* is designed to be *transparent* to the user. That is, the users doesn't know which services are local and which are remote.

local loop

The connection between a telephone company *Central Office* and a service subscriber.

local station

The network node to which the user is attached. See also: remote station.

LocalTalk

The *AppleTalk* physical *medium* used to connect Apple Computer products.

LocalTalk Link Access Protocol (LLAP)

The AppleTalk *datalink-layer* protocol used over *LocalTalk* networks. See also: *AppleTalk Remote Access Protocol, EtherTalk Link Access Protocol, TokenTalk Link Access Protocol.*

lockstep

A condition in which two hosts are alternating packet transmissions. There are some *protocols* which are designed to lockstep (e.g., *TFTP*). However, a protocol like *TCP* should allow multiple packets to be transmitted before requiring an *acknowledgment*; therefore, being forced to wait for an acknowledgment between each packet would be inefficient.

Logical Link Control (LLC)

The upper portion of the *datalink layer*, as defined in IEEE *802.2*. The LLC sublayer presents a uniform interface to the user of the datalink service, usually the *network layer*. Beneath the LLC sublayer is the *MAC* sublayer. See also: *802.x*.

Logical Unit (LU)

An *SNA* port through which a user gains access to network services. See also: *Physical Unit.*

long-haul

A long distance (*interLATA*) telephone circuit.

loopback

At the *physical layer*, loopback refers to a diagnostic procedure in which transmitted data is returned to the transmitter and compared with the original transmission. At the *network layer*, loopback refers to a packet sent by one application to another application on the same network node.

Low Entry Networking (LEN)

A *peer*-oriented, *SNA* extension which allows for easier construction and management of networks.

Low Frequency (LF)

A portion of the *electromagentic spectrum* with frequencies ranging from 30KHz to 300KHz.

LSB

See: *Least Significant Bit/Byte*

LSI

Large Scale Integration

LU

See: *Logical Unit*

LU 6.2

See: *Advanced Program-to-Program Communication*

lurking

No active participation on the part of a subscriber to a *mailing list* or *newsgroup*. A person who is lurking is just listening to the discussion. Lurking is encouraged for beginners who need to get up to speed on the history of the group. See also: *Electronic Mail, Usenet.*

M

M

See: *Mega-*

M bit

The More Data indication in *X.25* packets. It allows the sender to indicate a sequence of packets.

MAC

See: *Media Access Control*

MAC address

The *hardware address* of a device connected to a shared media. See also: *Media Access Control, Ethernet, token bus, token ring.*

Magnetic Ink Character Recognition (MICR)

A technique whereby printed characters can be read by a scanner and converted into textual data. The ink used in the process contains traces of a magnetic material which the scanner recognizes.

mail bridge

A *mail gateway* which forwards *Electronic Mail* between two or more networks while ensuring that the messages it *forwards* meet certain administrative criteria. A mail bridge is simply a specialized form of mail gateway which enforces an administrative policy with regard to what mail it forwards.

Mail Exchange Record (MX Record)

A *DNS* resource record type indicating which host can handle *Electronic Mail* for a particular domain.

mail exploder

Part of an *Electronic Mail* delivery system which allows a message to be delivered to a list of addresses. Mail exploders are used to implement *mailing lists.* Users send messages to a single address and the mail exploder takes care of delivery to the individual mailboxes in the list. See also: *email address.*

mail gateway

A machine which connects two or more *Electronic Mail* systems (including dissimilar mail systems) and transfers messages between them. Sometimes the mapping and translation can be quite complex, and it generally requires a *store-and-forward* scheme whereby the message is received from one system completely before it is transmitted to the next system, after suitable translations.

mail path

A series of machine names used to direct *Electronic Mail* from one user to another. This system of *email addressing* has been used primarily in *UUCP* networks which are trying to eliminate its use altogether. See also: *bang path.*

mail reflector

Part of an *Electronic Mail* delivery system which accepts a message and forwards it to another *email address.* Mail reflectors typically are used to make the location of a mailing list *transparent,* so that users who send *email* to that list don't need to know if it moves.

mail server

A software program which distributes files or information in response to requests sent via *email. Internet* examples include Almanac and netlib. Mail servers have also been used in *Bitnet* to provide *FTP*-like services.

mailing list

A list of *email addresses,* used by a *mail exploder,* to forward messages to groups of people. Generally, a mailing list is used to discuss certain set of topics, and different mailing lists discuss different topics. A mailing list may be moderated. This means that messages sent to the list are actually sent to a moderator who determines whether or not to send the messages on to everyone else. Requests to subscribe to, or unsubscribe from, a mailing list should ALWAYS be sent to the list's "-request" address (e.g., ietf-request@cnri.reston.va.us for the *IETF* mailing list). See also: *Electronic Mail, mail exploder.*

MAN

See: *Metropolitan Area Network*

Management Information Base (MIB)

The set of parameters an *SNMP* management station can query or set in the SNMP agent of a network device (e.g., router). Standard, minimal MIBs have been defined, and vendors often have Private Enterprise MIBs. In theory, any SNMP manager can talk to any SNMP agent with a properly defined MIB. See also: *CMIP, CMOT.*

Manchester encoding

The signaling method used by *CSMA/CA* and *CSMA/CD* transmission media. Each bit time is divided into two halves by a transition between positive and negative voltages in the middle of the bit time. A negative to positive transition indicates a 1-bit; a positive to negative transition indicates a 0-bit. Since a transition always occurs, Manchester encoding may be used by the receiving device to stay in synchronization with the transmitting device. See also: *electrical signaling* (includes figure).

Manufacturing Automation Protocol (MAP)

A subset of *OSI* standards, originated by General Motors, designed to maximize interoperability in areas where plain OSI standards are ambiguous or allow excessive options. See also: *Government OSI Profile, Technical and Office Protocols.*

MAP

See: *Manufacturing Automation Protocol*

mapping

The association of one set of values with another. For example, *hardware addresses* and *network-layer addresses*, or names and addresses.

Martian

A humorous term applied to packets that turn up unexpectedly on the wrong network because of incorrect *routing* entries. Also used as a name for a packet which has an altogether bogus (non-registered or malformed) *internet address.* See also: *bogon, kiss-of-death packet.*

maser

A device which generates electromagnetic signals in the microwave range. MASER is an acronym for Microwave Amplification by Stimulated Emission of Radiation.

MAU

See: *Multistation Access Unit*

Maximum Packet Lifetime (MPL)

The maximum amount of time an *AppleTalk* packet is allowed to exist on an network. See also: *Time To Live.*

Maximum Transmission Unit (MTU)

The largest *frame* length which may be sent on a physical *medium.* See also: *fragmentation.*

Mb/s

Megabits per second

Mean Time Between Failures (MTBF)

The average amount of time between failures of a device or system. See also: *Mean Time To Repair, fault tolerant.*

Mean Time To Repair (MTTR)

The average amount of time it takes to repair a failed device or system. See also: *Mean Time Between Failures, fault tolerant.*

Media Access Control (MAC)

The lower portion of the *datalink layer.* The MAC differs for various physical media. See also: *MAC Address, 802.3, Ethernet, Logical Link Control, token bus, token ring.*

medium

The physical cable to which nodes are attached, and over which signals are transmitted. In the case of *open-air transmissions,* the medium is air. *Radio Frequency* transmissions use no medium at all. See also: *coaxial cable, multi-mode fiber, single-mode fiber, twinaxial cable, Twisted-Pair, Unshielded Twisted-Pair.*

Medium Frequency (MF)

A portion of the *electromagnetic spectrum* with frequencies ranging from 300KHz to 3MHz.

Mega- (M)

In communications (i.e., Mb/s) and computer memory, 1M equals 2^{20} (1024^2 = 1,048,576). In other applications, 1M equals 10^6 (1,000,000).

mesh network

A network *topology* in which each node is connected to four adjacent nodes (*neighbors*). Edge nodes only have three neighbors, and corner nodes only have two neighbors.

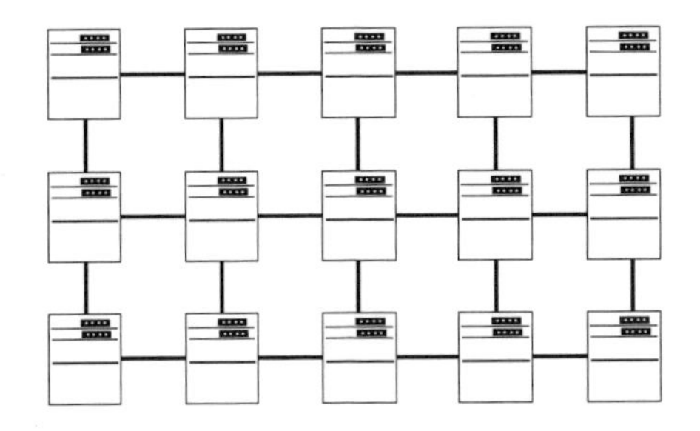

message

A self-contained unit of information, meaningful between two *peers.* At each *layer* of a *protocol stack,* a message may have a name which distinguishes it from messages at other layers. For example, a *datalink-layer* message is a *frame,* and a *network-layer* message is a *datagram.*

Message Handling System (MHS)

Specified in the *CCITT X.400* series of recommendations, MHS is the system of message *User Agents, Message Transfer Agents,* stores, and access units which provide *OSI*'s *Electronic Mail.*

message switching

See: *packet switching*

Message Transfer Agent (MTA)

An *OSI* application process which stores and forwards *X.400* messages. See also: *store-and-forward, User Agent.*

Message Telephone Service (MTS)

The official designation for tariffed, long-distance telephone service.

Metropolitan Area Network (MAN)

A data network intended to serve an area approximating that of a large city. Such networks are being implemented by innovative techniques, such as running *fiber-optic* cables through subway tunnels. A popular examplc of a MAN is *SMDS*. See also: *Local Area Network, Wide Area Network.*

MHz

Megahertz (one million cycles per second)

MIB

See: *Management Information Base*

MICR

See: *Magnetic Ink Character Recognition*

microcode

A set of low-level instructions which typically contain the programming a device requires to load a larger, more complex program.

Microcom Networking Protocol (MNP)

A proprietary error-correction protocol used by *modems* operating with data rates above 2400b/s.

microwave

A portion of the *electromagnetic spectrum* above 890MHz. Microwave transmissions, depending on the frequency, can be used for line-of-sight transmission or *satellite communication.*

MIDI

See: *Musical Instrument Digital Interface*

mid-level network

Mid-level networks, previously known as "regionals," make up the second level of the *Internet* hierarchy. They are the *transit networks* which connect the *stub networks* to the *backbone* networks.

MIF

See: *Minimum Internetworking Functionality*

MILNET

The MILNET was originally part of the *ARPANET.* As the ARPANET became the *Internet,* the MILNET was partitioned to allow United States military sites to have access to a reliable, secure data network.

MIME

See: *Multipurpose Internet Mail Extensions*

Minimum Internetworking Functionality (MIF)

An *OSI* principle which calls for minimal complexity within a *LAN* node when that node is communicating with a node off the LAN.

MNP

See: *Microcom Networking Protocol*

modem

See: *modulator/demodulator*

modem eliminator

A device or cable which allows two *DTEs* to communicate directly. It simulates the existence of *modems* and the intervening analog communications network (i.e., telephone circuit). Especially when referring to a cable, a modem eliminator is also called a null modem.

moderator

A person, or small group of people, who manage moderated *mailing lists* and *newsgroups.* Moderators are responsible for determining which *email* submissions are passed on to list. See also: *Electronic Mail, Usenet.*

modulate

The *encoding* of an *analog* signal onto a *carrier* signal. See also: *Amplitude Modulation, Frequency Modulation, Phase Modulation, Pulse Width Modulation.*

modulator/demodulator

A device which permits *digital* data to be transmitted over an *analog* communications network (i.e., telephone circuit). See also: *coder/decoder.*

Most Significant Bit/Byte (MSB)

The left-most bit in a byte, or the left-most byte in a *big-endian* word. That is, the bit or byte which has the most value. See also: *Least Significant Bit/Byte.*

mount

The making of a portion of a *file server's* filesystem accessible to other hosts. See also: *Network File System, Remote File System.*

MS-DOS

Microsoft's *Disk Operating System* for *Personal Computers.* See also: *PC-DOS.*

MSB

See: *Most Significant Bit/Byte*

MTA

See: *Message Transfer Agent*

MTBF

See: *Mean Time Between Failures*

MTTR

See: *Mean Time To Repair*

MTU

See: *Maximum Transmission Unit*

MUD

See: *Multi-User Dungeon*

multi-mode

A *fiber-optic* cable capable of simultaneously carrying multiple *light signals* of differing frequencies or phases. See also: *single mode, optical fiber.*

multicast

A packet with a special destination address which multiple nodes on the network may be willing to receive. See also: *group address, broadcast.*

multidomain network

An *SNA* network consisting of multiple *System Services Control Points.*

multidrop

A network *topology* in which multiple devices share a common transmission media. See also: *broadcast medium, CSMA/CA, CSMA/CD.*

multihomed host

A *host* which has connection to multiple networks. The host may send and receive data over any of the links but will not *route* traffic for other nodes. See also: *router.*

Multiple Virtual Systems (MVS)

An IBM mainframe *Operating System.* A pejorative for MVS is "Man Verses Machine."

multiplexer

A device which combines multiple data channels onto a single data communications medium. The combination is done according to a protocol or known algorithm so that the individual data channels may be separated at the remote end by a demultiplexor. See also: *interleaving.*

multi-protocol router

Any *router* which supports multiple *network-layer* protocols, such as *AppleTalk, CLNP, IP* and *IPX.* The concept for the multi-protocol router was invented by Noel Chiappa, a long time contributor to *TCP/IP* design and development.

Multipurpose Internet Mail Extensions (MIME)

An extension to *Electronic Mail* which provides the ability to transfer non-textual data, such as graphics, audio and fax.

Multistation Access Unit (MAU)

A *concentrator* used in *Local Area Networks.* It allows multiple network nodes to access a *LAN* through a single device, as opposed to a separate network *tap* for each node.

Multi-User Dungeon (MUD)

Adventure, role playing games, or simulations played on the *Internet.* Devotees call them "text-based virtual reality adventures." The games can feature fantasy combat, booby traps and magic. Players interact in *real-time* and can change the "world" in the game as they play it. Most MUDs are based on the *Telnet* protocol.

Musical Instrument Digital Interface (MIDI)

A standard *interface* between electronic musical instruments (e.g., keyboards, guitars), microphones, synthesizers, and computers. It allows *dig-*

itized music to be recorded by computer, where it may be modified, mixed and stored.

mux

See: *multiplexor*

MVS

See: *Multiple Virtual Systems*

MX Record

See: *Mail Exchange Record*

N

NAK

See: *Negative Acknowledgment*

Name Binding Protocol (NBP)

An AppleTalk *transport-layer* protocol which translates between user-defined *socket* names and their corresponding addresses. See also: *name resolution.*

name resolution

The process of *mapping* a name into its corresponding address. See also: *Domain Name System.*

named pipes

An *Application Program Interface* which extends the basic function of a *pipe* by allowing bi-directional communication and the ability to create a pipe to an application on a remote host.

namespace

A commonly distributed set of names in which all of the names are unique.

NAPLPS

See: *North American Presentation Level Protocol Syntax*

National Cable Television Association (NCTA)

A leading trade organization which represents United States cable television carriers.

National Exchange Carrier Association (NECA)

An association of local exchange carriers. The creation of NECA was mandated by the *FCC* after the AT&T divestiture.

National Information Infrastructure (NII)

The foundation for support of the *Internet.* This includes support for research and development, *Network Information Centers,* and other operations which are necessary to the use, if not the functioning, of a global *internet.* See also: *National Research and Education Network.*

National Institute of Standards and Technology (NIST)

United States governmental body which provides assistance in developing *standards.* NIST was formerly known as the National Bureau of Standards.

National Public Telecomputing Network (NPTN)

An organization, based in Cleveland, Ohio, devoted to making computer telecommunications and networking services as freely available as public libraries

National Research and Education Network (NREN)

The realization of an interconnected, gigabit computer network devoted to *High Performance Computing and Communications.* See also: *IINREN.*

National Science Foundation (NSF)

A United States government agency whose purpose is to promote the advancement of science. NSF funds science researchers, scientific projects, and infrastructure to improve the quality of scientific research. See also: *NSFNET.*

National Telecommunications and Information Administration (NTIA)

An agency of the United States Department of Commerce which is involved in the development of communication *standards.*

National Television System Committee signal (NTSC)

The *standard* specifying the format of television signals in the United States.

NAU

See: *Network Addressable Unit*

NBP

See: *Name Binding Protocol*

NBS
National Bureau of Standards (see: National Institute for Standards and Technologies)

NCCF
See: *Network Communications Control Facility*

NCP
See: *Network Control Program*

NCTA
See: *National Cable Television Association*

NCTE
See: *Network Channel Terminating Equipment*

NDIS
See: *Network Device Interface Specification*

NECA
See: *National Exchange Carrier Association*

Negative Acknowledgment (NAK)
The response to the receipt of a corrupted packet. See also: Acknowledgment.

neighbor
Any adjacent node on a network. Two nodes which are directly connected, by a point-to-point line or on a shared media, are said to be neighbors of each other. See also: *downstream neighbor, upstream neighbor.*

NetBIOS
See: Network Basic Input Output System

netiquette
A pun on "etiquette" referring to proper behavior on a network, *mailing list* or *newsgroup*. Proper behavior consists of adherence to written rules, called *Acceptable Use Policies*, and to unwritten understandings between long time users of the network, mailing list, or newsgroup. In the latter case, it is sometime difficult to determine what the understandings are until you trip over them; but, be assured that if you do, you will hear about it.

Netnews

See: *Usenet*

NetView

IBM's *network management* product. While its primary function is to manage *SNA* networks, it can also be used to collect information from non-*SNA* network components.

NetWare

Novell's network operating system. It is based on the *IPX protocol stack.*

network

A computer network is a data communications system which interconnects computer systems at various different sites. A network may be composed of any combination of *LANs*, *MANs* or *WANs*. See also: *backbone, transit network, stub network, internet.*

network address

The network portion of an *IP address.* For a class A network, the *network address* is the first byte of the IP address. For a class B network, the network address is the first two bytes of the IP address. For a class C network, the network address is the first three bytes of the IP address. In each case, the remainder is the host address. In the *Internet,* assigned network addresses are globally unique. See also: *subnet address, Internet Registry.*

Network Addressable Unit (NAU)

An host-based *Logical Unit, Physical Unit,* or *System Services Control Point* which is the source or destination of information transmitted by the *path-control layer* of an *SNA* network.

network architecture

A set of principles used as a basis for the design and implementation of a network. The principles cover *protocols, interfaces* and *message* formats. See also: *architecture, layered architecture, hierarchical architecture.*

Network Basic Input Output System (NetBIOS)

The standard *Personal Computer* interface to a data communication network.

network broadcast

A *broadcast* packet which is delivered to all nodes on the specified network. In *IP,* for example, a packet addressed to 132.245.255.255 would be delivered to all *nodes* on all *subnets* of the 132.245.0.0 network. Where necessary, *routers* would *forward* the packet to the network. See also: *directed broadcast, subnet broadcast, multicast.*

network byte order

The *byte order* of data as it is transmitted over a network. See also: *host byte order, big-endian, little-endian.*

Network Channel Terminating Equipment (NCTE)

Customer Premises Equipment which terminates telephone circuits.

Network Communications Control Facility (NCCF)

An IBM product through which users, and other programs, can monitor and manage network operation.

Network Control Program (NCP)

An *SNA,* host-generated program which controls an IBM communications controller (e.g., IBM 3725).

Network Device Interface Specification (NDIS)

An *datalink-layer* interface specification created by Microsoft and 3Com. On *PCs,* it provides a standard software *interface* to the *device driver* for the network interface card. See also: *Open Datalink Interface, Packet Driver Interface.*

Network File System (NFS)

A *protocol* developed by Sun Microsystems which allows a computer system to access files over a network as if they were on its local disks. This protocol has been incorporated in products by more than two hundred companies, and is now a *de facto standard* on the Internet. See also: *Remote File System.*

Network Information Center (NIC)

A provider of information, assistance and services to network users. See also: *InterNIC, Network Operations Center.*

Network Information Services (NIS)

A set of services, generally provided by a *NIC,* to assist users in using the network. Also, a set of services (previously known as Yellow Pages) used to maintain system files on *UNIX* hosts.

network layer

The third layer of the *OSI reference model.* This layer is responsible for *routing* data between nodes on the network. *IP* and *CLNP* are examples of network-layer protocols.

network-layer address

The identifier associated with the *network layer* of a node on a network. See also: *internet address, IPX address, OSI network address.*

network management

The remote monitoring and administration of network *entities* (i.e., *agents*) through the use of special *protocols* (e.g., *CMIP, SNMP*).

Network Management System (NMS)

The system which manages a *network* or *internet* by communicating with network management *agents*, which reside in various *nodes* and devices on the network. See also: *Common Management Information Protocol, Simple Network Management Protocol.*

Network News Transfer Protocol (NNTP)

A *protocol* for the distribution, inquiry, retrieval, and posting of news articles over a data communications network. See also: *Usenet.*

network number

The *network-layer* identifier for a logical network. See also: *network address, network-layer address, IPX address, OSI network address.*

Network Operating System (NOS)

The software which manages network resources for a node on a network. These resources may include *Electronic Mail, file servers* and *print servers.* A NOS may also provide security and access control. See also: *Operating System.*

Network Operations Center (NOC)

A location from which the operation of a *network* or *internet* is monitored. Additionally, this center usually serves as a clearinghouse for connectivity problems and efforts to resolve those problems. See also: *Network Information Center.*

Network Service Access Point (NSAP)

An identifier for an *OSI* network service. NSAPs are encoded within *OSI network addresses.*

Network Terminal Option (NTO)

An IBM product which enables an *SNA* network to accommodate a select group of non-SNA devices.

Network Time Protocol (NTP)

A protocol which assures accurate local timekeeping with reference to radio and atomic clocks located on the *Internet.* This protocol is capable of synchronizing distributed clocks within milliseconds over long periods of time.

Network Visible Entity (NVE)

In *Appletalk,* any *service* which may be accessed by a *client.* Generically, any device or *node* which can be accessed by other nodes on the same *internet.*

networking

The use of communications media (e.g., *Local Area Networks, leased lines*) to interconnect computers so that users on those computers may share resources (e.g., processing power, disk space, I/O devices).

NFS

See: *Network File System*

nibble

The humorous term for a 4-bit unit of information, which holds one *hexadecimal* "digit." The term comes from the fact that a nibble is half of a *byte.*

NIC

See: *Network Information Center*

NII

See: *National Information Infrastructure*

NIS

See: *Network Information Services*

NIST

See: *National Institute of Standards and Technology*

NMS

See: *Network Management System*

NNTP

See: *Network News Transfer Protocol*

NOC

See: *Network Operations Center*

Nodal Switching System (NSS)

The main routing nodes in the *NSFNET backbone.* See also: *National Science Foundation.*

node

An *network-layer* addressable device attached to a computer data network. See also: *End System, host, Intermediate System, router.*

node type

An *SNA* classification for network nodes based on the protocols they support and the *Network Addressable Units* they could contain. Peripheral nodes are types 1 and 2; subarea nodes are types 4 and 5.

Noelgram

Any lengthy, detailed *Electronic Mail* message sent to a *mailing list* or *newsgroup.* The term comes from the lengthy, stylized messages drafted by Noel Chiappa, a long time participant in the design of *TCP/IP,* and the inventor of the *multi-protocol router.* See also: *flame.*

non-blocking

Descriptive of a device or protocol through which a *message* path always exists. That is, high volume traffic cannot create a busy condition.

Non-return to Zero (NRZ)

An signaling method which represents binary data with opposite, alternating high and low voltages. During data transmission, there is no signal with a zero voltage. See also: *electrical signaling* (includes figure)

North American Presentation Level Protocol Syntax (NAPLPS)

A screen format and videotex graphics protocol developed by AT&T. It is based on Canada's Telidon videotex-graphics protocol and standardized within *ANSI.*

NOS

See: *Network Operating System*

NPTN

See: *National Public Telecomputing Network*

NREN

See: *National Research and Education Network*

NRZ

See: *Non-return to Zero*

NSAP

See: *Network Service Access Point*

NSF

See: *National Science Foundation*

NSFNET

The *NSFNET,* funded by *NSF,* is an essential part of academic and research communications in the United States. It is a high-speed "network of networks" which is hierarchical in nature. The NSFNET *backbone* is a network currently composed of 16 nodes, interconnected with *T3* service, spanning the continental United States. Attached to the backbone are *mid-level networks,* to which campus and local networks are attached. *NSFNET* also has connections out of the United States to Canada, Mexico, Europe, and the Pacific Rim.

NSS

See: *Nodal Switching System*

NTIA

See: *National Telecommunications and Information Administration*

NTO

See: *Network Terminal Option*

NTP

See: *Network Time Protocol*

NTSC signal

See: *National Television System Committee signal*

null character

A *control character* which may be inserted into or removed from a data *message* without affecting the meaning of the message. Null characters are often called nulls, and typically have a hexadecimal value of 00h.

null modem

See: *modem eliminator*

NVE

See: *Network Visible Entity*

NYNEX

A *Regional Bell Operating Company* which services the New York / New England region of the United States.

O

OCLC

See: *On-line Computer Library Catalog*

OCR

See: *Optical Character Recognition*

octal

The base-8 numbering system. It is used for computer notation because 8 values can be represented by 3 *bits*. It is more popular on 9-bit hardware architectures. See also: *binary, hexadecimal.*

octet

An 8-bit unit of information. This term is used in networking, rather than byte, because some systems have bytes that are not 8 bits long. Also, an octet is not necessarily 8-bit aligned.

ODA

See: *Office Document Architecture*

odd parity

See: *parity check*

ODI

See: *Open Datalink Interface*

OEM

See: *Original Equipment Manufacturer*

OEM-I

The standard *interface* to IBM mainframes. It is a bus and tag interface in which signals from the *CPU* go out over the bus cable and signals from the *peripheral devices* return over the tag cable.

off-hook

A condition indicating that a telephone circuit is active. Literally, "the telephone is off the hook" (from the days when the telephone handset rested in a hook when not in use). See also: *on-hook.*

off-line

A condition in which a device is not available for use. See also: *on- line.*

offered load

The amount of data, measured in *octets* or *packets,* being injected into a network. See also: *carrying capacity.*

Office Document Architecture (ODA)

A *ISO* standard document format and handling system which allows documents to be shared by otherwise incompatible systems.

OIW

See: *Workshop for Implementators of OSI*

on-hook

A condition indicating that a telephone circuit is inactive (hung up). See also: *off-hook.*

on-line

A condition is which a device is available for use. See also: *off-line.*

ONC

See: *Open Network Computing*

ones density

The requirement, for *digital* circuits, that eight consecutive 0-bits cannot be sent as part of a data stream. The requirement stems from the fact that some telephony equipment depends on bit transitions to maintain time synchronization. The insertion of a 1-bit for this reason is called *zero code suppression.* See also: *bit stuffing.*

On-line Computer Library Catalog (OCLC)

A nonprofit membership organization offering computer-based services to libraries, educational organizations, and their users. The OCLC library information network connects more than 10,000 libraries worldwide. Libraries use the OCLC System for cataloging, interlibrary loan, collection development, bibliographic verification, and reference searching.

open-air transmission

A communication link which does not use a physical *medium*, such as cable or fiber. Open-air transmission may use *Radio Frequency* (shortwave, microwave, etc.), *infrared*, or *laser*, and are generally limited to short distances and line-of-sight use.

Open Datalink Interface (ODI)

An *datalink-layer* interface specification created by Novell. On *PCs*, it provides a standard software *interface* to the *device driver* for the network interface card. See also: *Network Device Interface Specification, Packet Driver Interface.*

Open Network Computing (ONC)

A distributed applications *architecture* controlled and promoted by a Sun Microsystems consortium.

Open Shortest-Path First Interior Gateway Protocol (OSPF)

A *Link-State*, as opposed to *Distance Vector*, *routing protocol*. It is an *Internet* Standard *Interior Gateway Protocol*. See also: *Routing Information Protocol, Exterior Gateway Protocol.*

Open Software Foundation (OSF)

A consortium of computer vendors jointly developing a standard version of *UNIX*.

Open Systems Interconnection (OSI)

A suite of protocols, designed by *ISO* committees, to be the international standard computer *network architecture*. See also: *OSI reference model.*

Operating System (OS)

The software which controls the execution of programs on a computer. An OS typically provides process management, timer management, and file management for the applications. In this context, applications include *device drivers*, *protocol stacks*, and the software dependent upon them. See also: *Disk Operating System, Network Operating System.*

Optical Character Recognition (OCR)

A technique by which standard printed material, as well as graphics and pictures, may be converted into digital information suitable for transmission or storage. See also: *digitize.*

optical fiber

A plastic or glass transmission *medium* which is used to carry *light signals.* See also: *multi-mode, single mode, laser, Light Emitting Diode.*

Original Equipment Manufacturer (OEM)

The manufacturer of equipment which is marketed by another vendor, usually in conjunction with that vendor's own equipment.

OS

See: *Operating System*

OS/2

An operating system developed by IBM and Microsoft for use on *Personal Computers.* OS/1 was the name of an IBM mainframe operating system which has long since been replaced by *MVS.*

OSF

See: *Open Software Foundation*

OSI

See: *Open Systems Interconnection*

OSI network address

The identifier used to locate an *OSI* transport entity. It is a formatted, variable length address which may be up to 20 octets long. See also: *Authority and Network Identifier.*

OSI presentation address

The identifier used to locate an *OSI* application entity. It is composed of an *OSI network address* and up to three selectors. A selector may be present for the transport, session or presentation entities associated with the application entity.

OSI reference model

A seven-layered structure designed to describe computer *network architectures* and the ways in which data passes through them. This model was developed by the ISO in 1978 to clearly define the *interfaces* and *protocols* for multi-vendor networks, and to provide users of those networks with conceptual guidelines in the construction of such networks. Each of the layers shown in the diagram are described in this glossary.

Layer 7	Application
Layer 6	Presentation
Layer 5	Session
Layer 4	Transport
Layer 3	Network
Layer 2	Datalink
Layer 1	Physical

OSInet

A test network which provides *OSI* product vendors with a environment suitable for *interoperability* testing. It is sponsored by the *National Institute of Standards and Technologies.*

OSPF

See: *Open Shortest-Path First Interior Gateway Protocol*

out-of-band signaling

The passing of control information on a different channel than the data. For example, *B-ISDN* has a 16Kb/s control channel in addition to two data channels. See also: *in-band signaling.*

overhead

All non-userdata information which is transmitted over a communication channel. This includes *headers, trailers, control characters,* and *messages* for ancillary protocols (e.g., *routing, network management*).

overrun

Data which is lost when a receiver cannot accept data at the rate at which it is being transmitted. See also: *underrun.*

P

PABX

See: *Private Automatic Branch Exchange*

Pacific Bell

The *Regional Bell Operating Company* which services the west-coast region of the United States.

pacing group

The number of *Path Information Units* which may be sent in an *SNA* session before a response is received. See also: *flow control, sliding window.*

packet

A unit of data sent across a network. "Packet" a generic term used to describe units of data at all layers of the protocol stack, but it is most correctly used to describe application data units. See also: *datagram, frame, Protocol Data Unit.*

Packet Assembler/Disassembler (PAD)

A *protocol converter* which allows devices not equipped for packet switching to communicate over an *X.25* network.

Packet Authentication Protocol (PAP)

A *PPP* authentication protocol. It is a basic password exchange protocol and has been superceded by the *Challenge Handshake Authentication Protocol.*

Packet Driver Interface (PDI)

An *datalink-layer* interface specification created by FTP Software. On *PCs*, it provides a standard software *interface* to the *device driver* for the network interface card. PDI was the first such specification. See also: *Network Device Interface Specification, Open Datalink Interface.*

Packet Internet Groper (PING)

A program used to test reachability of a destination by sending it an echo *request* and waiting for a *reply*. The term is also used as a verb: "Ping host X to see if it is up!" See also: *probe, AppleTalk Echo Protocol, Internet Control Message Protocol.*

packet radio

A network in which nodes *broadcast* packets to each other over common *Radio Frequencies*. See also: *Aloha.*

Packet Switch Node (PSN)

A dedicated computer whose purpose is to accept, *route* and *forward* packets in a packet switched network. See also: *packet switching, router.*

packet switching

A communications paradigm in which *packets* (*messages*) are individually *route*d between hosts, with no previously established communication *path*. See also: *circuit switching, connection-oriented, connectionless.*

PAD

See: *Packet Assembler/Disassembler*

pad characters

In *synchronous* transmission, *SYN* characters which are inserted at the beginning and end of a *message* to ensure proper reception of the message. In *CSMA/CA* and *CSMA/CD* transmission, null characters which are added to end of a short message to bring the frame length up to the minimum required for the media

Palo Alto Research Center (PARC)

The research and development arm of Xerox. From the early 1970's through the mid-1980's, PARC invented such things as the *WIMP interface*, the laser printer, and the *Local Area Network.*

PAP

See: *Packet Authentication Protocol, Printer Access Protocol*

parallel sessions

Multiple concurrent *SNA* sessions executing within the same processor.

parallel interface

A network *interface* which transmits each of the bits in a character simultaneously. This requires multiconductor cable of the use of multiple frequencies over a single communications channel. See also: *serial interface.*

PARC

See: *Palo Alto Research Center*

PARC Universal Packet

Originally the name of a packet *format*, PUP became the name for an internet *architecture* and suite of *protocols* developed by Xerox. PUP is not widely used today. See also: *Palo Alto Research Center.*

parity bit

A additional bit which is transmitted with each character and is used to force an even or odd number of 1-bits per character. See also: *parity check.*

parity check

A simple error checking mechanism which can detect the transposition of an odd number of bits per character by counting the number of 1-bits in the character and parity bit, then determining if that count is odd, for odd parity, or even, for even parity. See also: *parity bit.*

passive device

A device which does not require external power to operate. It uses *phantom power* to perform its functions, which are typically limited compared to an *active device.* Some *transceivers* and *wiring hubs* are passive devices. Standard telephones are also passive devices.

path

The sequence of *routers* and *links* through which a packet passes on the trip from the packet's originating node to its destination node. See also: *bang path, mail path.*

path-control layer

The *network-layer* protocol in the *SNA* protocol stack. It determines the *routing* for packets in the network. See also: *path.*

Path Information Unit (PIU)

An *SNA* packet.

PBX
See: *Private Branch Exchange*

PC
See: *Personal Computer*

PC-DOS
IBM's *Disk Operating System* for *Personal Computers.* See also: *MS-DOS.*

PCI
See: *Protocol Control Information*

PCM
See: *Pulse Code Modulation*

PCMCIA
See: *Personal Computer Memory Card Industry Association*

PCD
See: *Personal Communication Device*

PDM
Pulse Duration Modulation (see: *Pulse Width Modulation*)

PCS
See: *Personal Communication Service*

PD
Public Domain

PDI
See: *Packet Driver Interface*

PDN
See: *Public Data Network*

PDU
See: *Protocol Data Unit*

peer (hardware)
A *node* on a *network* which is an equal to other nodes on that network. For example, all nodes on a *Local Area Network* are peers.

peer (software)

A *process* which is communicating to another process, at the same *layer* in the *protocol stack*, on another *node*. In the figure below, processes which communicate across a dotted line are peers. If the processes are *application* processes, for example, they are said to be *application-layer* peers. See also: *OSI reference model*

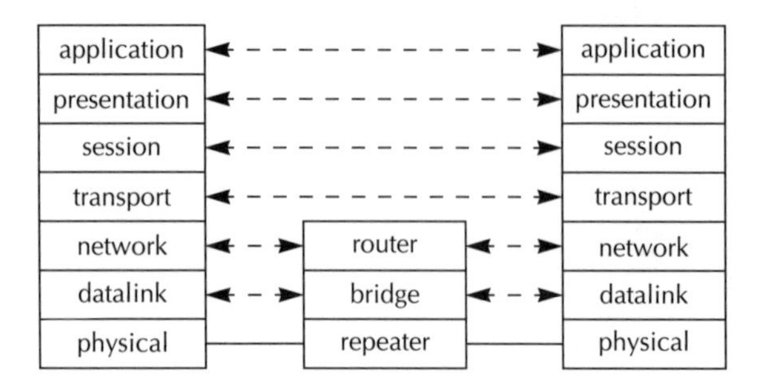

PEM

See: *Privacy Enhanced Mail*

peripheral device

A device which is not central to the operation of a computer system or network. Printers, magnetic tape drives, and terminals are examples of peripheral devices.

Permanent Virtual Circuit (PVC)

An *X.25* connection which remains open, and is dedicated to the end points of the connection. See also: *Switched Virtual Circuit.*

Personal Communication Device (PCD)

A communication device which uses *PCS* for voice and data communication.

Personal Communication Service (PCS)

An emerging technology which is a digital version of *cellular radio*. PCS technology allows many more connections using the same amount of *bandwidth*.

Personal Computer (PC)

The generic term for a single-user, microprocessor based computer which has an *architecture* derived from the original IBM Personal Computer. In an abstract sense, Apple Computer's Macintosh is a personal computer, but it is not a *Personal Computer* in that is based on a different architecture.

Personal Computer Memory Card Industry Association (PCMCIA)
The creator of an emerging standard for a small computer (i.e., laptop) *interface.* Version 1 of the standard defined an interface for memory expansion and software expansion cards. Version 2 of the interface allows many devices (e.g., *modems*) to use the PCMCIA interface.

phantom power
Electrical energy, required to run a *passive device,* stolen from the data communications circuit. See also: *active device.*

Phase Modulation (PM)
A transmission method in which data is *encoded* over a *carrier* signal by variations in the phase angle of the carrier. See also: *Amplitude Modulation, Frequency Modulation, Pulse Width Modulation.*

physical address
See: *hardware address*

physical layer
The lowest (first) layer of the *OSI reference model.* It is responsible for the manipulation of bits on the physical media. The media may be electrical (e.g., *coaxial* or *Twisted-Pair*), etheric (e.g., microwave), or optical (e.g., *single-mode* or *multi-mode* fiber).

Physical Unit (PU)
An *SNA* device which monitors and manages a node's resources. PUs have the same classification levels as nodes. See also: *Logical Units, node types.*

picture element (pixel)
The smallest unit of a picture, or a graphics or video display. The characteristics of a pixel (grey scale or RGB intensities) can be *digitized* for transmission or storage.

piggyback
The carrying of control information in data packets in order to reduce *overhead.* For example, *TCP* piggybacks *acknowledgments* on data *segments*

to avoid the necessity of sending a separate packet carrying only the acknowledgment.

PIN

See: *Positive, Intrinsic, Negative*

PING

See: *Packet Internet Groper*

pipe

A software mechanism which takes the *stream* output of one *application* and passes it to another application as input. See also: *named pipe.*

PIU

See: *Path Information Unit*

pixel

See: *picture element*

Point Of Presence (POP)

A site, on the subscribers premises, where there exists a collection of telecommunications equipment, usually digital *leased lines* and *multi-protocol routers.* See also: *Customer Premises Equipment.*

point-to-point

A network segment which directly connects exactly two network nodes. There are no intermediate processors above the physical layer; however, there might be telephone switching facilities.

Point-to-Point Protocol (PPP)

A protocol which provides a method for transmitting packets over *serial, point-to-point* links. It was originally developed for *IP,* but specifications for *AppleTalk* and *IPX* have been developed. See also: *Serial Line IP.*

poison reverse

A method used to reduce *convergence time* in some *routing protocols* which use *Distance-Vector algorithms.* Poison reverse causes *routes* learned over a particular interface to be advertised over that interface with a metric of infinity (i.e., one higher than the maximum metric permitted by the protocol).

polling

A control method in which a central, master station systematically queries subsidiary (slave) stations to determine if they require servicing.

POP

See: *Point Of Presence, Post Office Protocol*

port

A *transport-layer* demultiplexing value used to distinguish among multiple *applications* running on the same *node*. See also: *ephemeral port, well-known port, socket, Transmission Control Protocol, User Datagram Protocol.*

POSI

See: *Promoting Conference for OSI*

Positive, Intrinsic, Negative (PIN)

A photodetector used to convert the energy in a *light signal* into an electrical signal. See also: *laser, Light Emitting Diode.*

POSIX

Portable Operating System Interface for Computer Environments is an operating system *Application Program Interface* developed by *IEEE* to increase the portability of *application* software. See also: *sockets interface.*

Post Office Protocol (POP)

A protocol designed to allow single user hosts to read *Electronic Mail* from a server. There are three versions: POP, POP2, and POP3. Latter versions are not compatible with earlier versions.

Postal Telegraph and Telephone (PTT)

Outside the United States, PTT refers to a telephone service provider, which is usually a monopoly, in a particular country.

postmaster

The person responsible for taking care of *Electronic Mail* problems, answering questions about users, and other related work at a site. Often, the postmaster work in a *NIC* or *NOC.*

PostScript (PS)

A semi-standardized document format language. It allows text and graphics to be stored in *ASCII* format, to facilitate transfer of the document. Most laser printers can handle PS documents.

PPP

See: *Point-to-Point Protocol*

preamble

A bit sequence which is transmitted prior to the start of a *frame*. It is used to indicate to the receivers that a frame is about to be transmitted. The receivers may use it to synchronize their link timing.

presentation layer

The sixth layer of the *OSI reference model*. This layer is responsible for data manipulation functions common to many applications. *Compression* and *encryption* are example of presentation-layer functions.

Primary-rate ISDN (P-ISDN)

An *ISDN* service providing 23 64Kb/s data/voice (B) channels and one 64Kb/s control (D) channel. This service is also called 23B+D. See also: *Basic-rate ISDN*.

print server

A dedicated network node which provides printing facilities for other nodes on the network. The facilities may include text, graphics or plotting capabilities.

printable characters

A *character* for which there is a textual representation (e.g., letters, numbers, punctuation). See also: *white space, control characters*.

Printer Access Protocol (PAP)

An AppleTalk *session-layer* protocol which. See also: *AppleTalk Data Stream Protocol, AppleTalk Session Protocol*.

Privacy Enhanced Mail (PEM)

Electronic Mail which provides confidentiality, *authentication* and message integrity using various *encryption* methods.

Private Automatic Branch Exchange (PABX)

A *PBX* which does not require operator intervention to place a call. See also: *Computerized Branch Exchange*.

Private Branch Exchange (PBX)

Originally an operator's switchboard, PBX has come to mean a telephone switch located on the subscribers premises which connects users to external telephone lines and to each other. Many features are offered on most PBXs (e.g., *camp-on, call-waiting, call-forwarding*). See also: *Computerized Branch Exchange, Private Automatic Branch Exchange.*

Private Management Domain (PRMD)

An *X.400 Message Handling System* used by private organizations for *Electronic Mail* delivery. See also: *Administration Management Domain.*

probe

A packet, sent by a one network node to another, requesting an *acknowledgment.* The probe packet itself serves to indicate to the receiver that the sender is there. See also: *keep-alive, Packet Internet Groper, tickle.*

process

A uniquely identifiable incarnation of a procedure. For example, *applications* running on *hosts* are processes; *routing protocols* running on *routers* are processes.

PROM

Programmable Read Only Memory

promiscuous mode

An operating method used on a *broadcast medium* wherein a node accepts all *frames,* not only *broadcasts* or frames addressed to it. Promiscuous mode is most often used by devices which analyze network traffic and, therefore, need to analyze every frame.

Promoting Conference for OSI (POSI)

Japan's *OSI* "800 pound gorilla." It is composed of executives from six major Japanese computer manufacturers and Nippon Telephone and Telegraph. It is responsible for policy setting and promotion of OSI.

propagation delay

The time it takes a signal to traverse a network segment, or the time it takes a signal to travel from one node to another. The delay exists because signals do not propagate instantaneously. Electrical signals on a wire travel at approximately 33% of *lightspeed. RF* signals in *coaxial cable* travel at approximately 90% of lightspeed. *Light signals* in an *optical fiber* travel at approximately 99% of lightspeed.

Prospero

A distributed filesystem which provides the user with the ability to create multiple views of a single collection of files distributed across the *Internet.* Prospero provides a file naming system, and file access is provided by existing access methods (e.g., *anonymous FTP* and *NFS*). The Prospero protocol is also used for communication between *clients* and *servers* in the *archie* system. See also: *archive site, Gopher, Wide Area Information Servers.*

protocol

A formal description of *message* formats and the rules two computers must follow to exchange those messages. Protocols can describe low-level details of machine-to-machine *interfaces* (e.g., the order in which bits and bytes are sent across a wire) or high-level exchanges between *application* programs (e.g., the way in which two programs transfer a file across an *internet*).

Protocol Control Information (PCI)

The protocol information (*header*) added to the *Service Data Unit* handed down from the layer above. The combination of PCI and *SDU* form the *Protocol Data Unit* which is then passed to the layer below.

protocol converter

A device or program which translates between different *protocols* which serve similar functions (e.g., *TCP* and *TP4*).

Protocol Data Unit (PDU)

"PDU" is internationalstandardscomitteespeak for *packet.* See also: *Protocol Control Information, Service Data Unit.*

protocol stack

A *layered* set of *protocols* which work together to provide a set of networking functions. See also: *layered architecture, hierarchical architecture.*

proxy

An operational paradigm wherein one system responds to protocol requests directed to another system. This mechanism is used in *network management* to avoid the necessity of implementing full protocol stacks in simple devices. See also: *proxy ARP.*

proxy ARP

The technique in which one network node, usually a *router*, answers *ARP* requests intended for another node. By "faking" its identity, the router accepts responsibility for *routing* packets to the "real" destination. Proxy ARP allows a site to use a single *IP address* with two physical networks. *Subnetting* would normally be a better solution.

PS

See: *PostScript*

PSN

See: *Packet Switch Node.*

PSTN

See: *Public Switched Telephone Network*

PTT

See: *Postal, Telegraph and Telephone*

Public Data Network (PDN)

A *packet-switched* data carrier. *TELENET* is an example of a PDN. See also: *Public Switched Telephone Network.*

public key

An *RSA encryption* key, known to all users, which is used to *encrypt* data for a specific individual.

Public Switched Telephone Network (PSTN)

The *dial-up* telephone network. See also: *Public Data Network.*

Public Utilities Commission (PUC)

A local or state government organization which oversees the actions of the public utility companies (e.g., cable, electric, gas, telephone).

PUC

See: *Public Utilities Commission*

Pulse Code Modulation (PCM)

A voice *digitizing* technique in which the *analog* voice signal is sampled, typically at 8KHz, and the amplitude of the signal at the sampling interval is *encoded* as a binary value, typically in eight bits.

Pulse Duration Modulation (PDM)

See: *Pulse Width Modulation*

Pulse Width Modulation (PWM)

A transmission method in which data is *encoded* over a *carrier* signal by variations in the width of carrier pulses. PWM is also known as Pulse Duration Modulation (PDM). See also: *Amplitude Modulation, Frequency Modulation, Phase Modulation.*

PUP

See: *PARC Universal Packet*

push-down stack

See: *stack*

Q

Q bit

A bit in an *X.25* packet which the *DTE* may use to indicate that it wishes to transmit data on more than one level.

QOS

See: *Quality Of Service*

Quality Of Service (QOS)

A parameter in many protocols which may be used to request special treatment for a *message* (e.g., high-reliability, low-delay). See also: *Type Of Service.*

query packet

A packet sent by an *application* seeking information, in the form of a *reply packet*, from another application. See also: *query/reply protocol, request packet.*

query/reply protocol

A communications paradigm in which one *application* seeks information from another in a two packet exchange (one packet in each direction). *Name resolution* protocols are examples of query/reply protocols. See also: *query packet, reply packet, request/response protocol.*

queue

A backlog of packets, maintained in the order in which they were received, awaiting processing or transmission. See also: *FIFO, stack.*

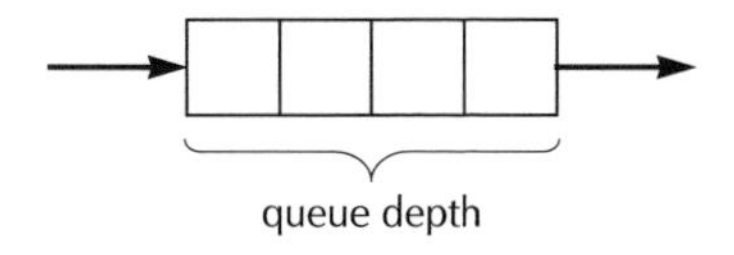

quiescent

A state in which voice or data are not being carried over a communications channel. It is typically a temporary *idle* period forced by the communication channel, as opposed there simply being no information to transmit.

quiet time

The amount of time a rebooted *host* should remain quiet (i.e., before it sends any packets). The time is dictated by the need for packets associated with the host's previous incarnation to expire on the network, so as not to be confused with packets sent by the new incarnation. See also: *Maximum Packet Lifetime, Time To Live.*

quipu

Part of *ISODE*'s *X.500* implementation. The name is taken from a device used by ancient Peruvians for calculating and record keeping. It was made from a main cord with smaller, multicolored cords attached and knotted.

R

Radio Frequency (RF)

The frequencies at which electromagnetic energy may be made to radiate coherently. This is typically all frequencies above 150KHz.

Radio Frequency Interference (RFI)

The *Radio Frequency* radiation which leaks from a device which has been linked to a transmission *medium.* See also: *Electromagnetic Interference.*

RARE

See: *Réseaux Associés pour la Recherche Européens*

RARP

See: *Reverse Address Resolution Protocol*

rate center

A geographic point used by telephone service providers as a standard from which *interLATA* mileage-rate distance measurements are made.

RBOC

See: *Regional Bell Operating Company*

RCP

See: *Remote copy program*

RDB

See: *Relational Database*

Read the F*cking Manual (RTFM)

This acronym is often used when someone asks a simple or common question, the answer to which is readily available.

real-time
An operating mode in which data is transmitted and processed at speeds sufficient to allow *interactive* control.

reassembly
The *IP* process in which a previously fragmented packet is reassembled before being passed to the *transport layer.* See also: *fragmentation.*

Receive Only (RO)
An operational mode in which a network device is capable of receiving data but cannot transmit. Typically, *print servers* are RO devices.

recursive
See: *recursive*

Red, Green, Blue (RGB)
The three phosphor colors used in color CRTs. By varying the intensity of each of these primary colors in combination, any secondary color may be produced. RGB is also the term used for the interface to RGB monitors.

redundancy
Extra information which is transmitted or stored against the possibility of essential information being lost. See also: *Forward Error Correction.*

regional
See: *mid-level network*

Regional Bell Operating Company (RBOC)
The local telephone company in each of the seven United States regions. See also: *Ameritech, Bell Atlantic, Bell South, NYNEX, Pacific Bell, Southwestern Bell, US West.*

Relational Database (RDB)
A database, which may be distributed, which operates by linking tables according to columns shared by those tables. It allows for simply configuration of the database and is very flexible. See also: *Structured Query Language.*

Reliable Transfer Service Element (RTSE)
A lightweight *OSI* application service used above *X.25* networks.

Remote File System (RFS)

A distributed file system developed by AT&T and distributed with *UNIX* System V. See also: *Network File System.*

Remote Job Entry (RJE)

The submission of data processing jobs over a network.

remote login

Operating on a remote computer, using a protocol over a computer network, as though locally attached. See also: *Telnet, Virtual Terminal Protocol.*

Remote Operations Service Element (ROSE)

A lightweight *RPC* used in *OSI* message handling, and *directory* and *network management applications.*

Remote Procedure Call (RPC)

An easy and popular paradigm for implementing the *client-server model* of distributed computing. In general, a request is sent to a remote system to execute a designated procedure, using arguments supplied, and the result returned to the caller. There are many variations and subtleties in various implementations, resulting in a variety of different (incompatible) RPC protocols.

remote station

The node at the other end of a network connection. See also: local station.

repeater

A device which propagates *digital* signals from one cable to another. The signal is amplified, cleaned, and retimed. See also: *bridge, gateway, router.*

reply packet

A packet sent in response to a *query packet.* It contains the information requested in the query packet. See also: *unsolicited reply, query/reply protocol, response packet.*

request packet

A packet, usually sent by a *client* to a *server,* which solicits an action and, usually, a *response packet.* See also: *query packet, request/response protocol.*

request/response protocol

A communication paradigm in which one *application,* usually a *client,* sends a *request packet* to another application, usually a *server,* which performs some action and returns a *response packet.* See also: *query/reply protocol.*

Request For Comments (RFC)
The documentation series, begun in 1969, which describes the *Internet* suite of protocols and related experiments. Not all (in fact, very few) RFCs describe Internet Standards, but all Internet Standards are written up as RFCs. The RFC series of documents is unusual in that the proposed protocols are forwarded by the Internet research and development community, acting on their own behalf, as opposed to the formally reviewed and standardized protocols that are promoted by organizations such as *ISO* and *ANSI.* See also: *For Your Information, STD.*

Request To Send (RTS)
A *modem* signal which indicates to the modem that the *DTE* has data to send. See also: *RS-232.*

Réseaux Associés pour la Recherche Européens (RARE)
An international organization founded in 1986 with the mission of promoting and participating in the creation of a high-quality computer communications *infrastructure* to support Europe's research community. Established to remove technical and organizational barriers between national networks in Europe, RARE currently has a membership including over 40 countries and organizations throughout Europe, India, Israel, the Republic of Korea, and South Africa. RARE also supports the *RIPE NCC.*

Réseaux IP Européens (RIPE)
A collaboration between European networks which use the *TCP/IP* protocol suite. It was formed in 1989 and provides the administration and technical coordination necessary to ensure interoperability between the member service providers.

response packet
A packet sent in reply to a *request packet.* See also: *unsolicited response, request/response protocol, reply packet.*

response time
The elapsed time between a user's entry and its response. In terminal sessions, this may be the time between hitting a key and seeing its echo,

or the time between the entry of a command and the beginning of its response. In Remote Job Entry, it is the latter of those two.

retransmission

A repeat transmission of a unit of data which was previously transmitted, typically because the transmitter never received an *acknowledgment* of the first transmission.

retransmissive star

A passive, *fiber-optic* component which accepts a *light signal* on an input *optical fiber* and splits the signal onto multiple fibers.

retry

In *Bisync*, a *retransmission* of the current data block. Generically, any retransmission caused by a time-out.

Return to Zero (RZ)

An signaling method which allows the line voltage to return to zero after each bit is transmitted. See also: *electrical signaling* (includes figure)

Reverse Address Resolution Protocol (RARP)

A protocol which provides the reverse function of *ARP.* RARP maps a *hardware address* to an *internet address.* It is used primarily by diskless nodes when they first initialize to find their internet address. See also: *BOOTP, Dynamic Host Configuration Protocol, MAC address.*

reverse channel

A low-*bandwidth* channel which the receiver of data may use to send control signals to the transmitter of the data. See also: *out-of-band signaling.*

Reverse Interrupt (RVI)

A *bisync control character* sequence sent by the receiver, to the transmitter, requesting termination of the transmission in progress.

RF

See: *Radio Frequency*

RFC

See: *Request For Comments*

RFC 822

An Internet Standard format for *Electronic Mail* message *headers.* Mail experts often refer to "822 messages." The name comes from RFC 822, which contains the specification. The 822 format was previously known as 733 format.

RFI

See: *Radio Frequency Interference*

RFS

See: *Remote File System*

RGB

See: *Red, Green, Blue*

RI

See: *Ring Indicator*

Ring Indicator (RI)

A *modem* signal which indicates that a remote modem has called (literally, "the phone is ringing"). See also: *RS-232.*

ring network

A network *topology* in which each node is connected to two adjacent nodes (*neighbors*). Ring networks have the advantage of not needing *routing* because all packets are simply passed to a node's *upstream neighbor.* See also: *802.5, token ring, star-wired ring network.*

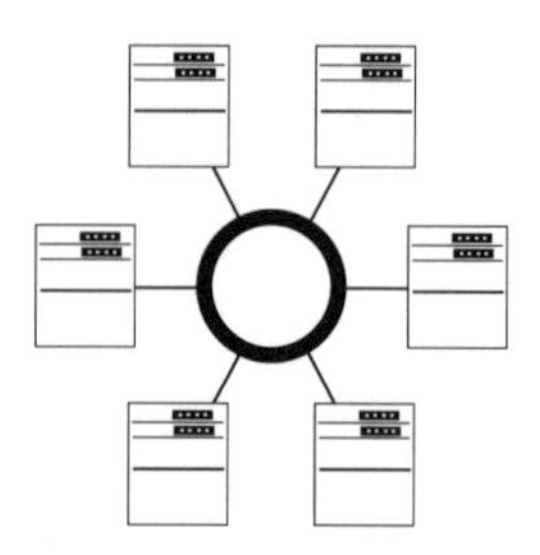

RIP

See: *Routing Information Protocol*

RIPE

See: *Réseaux Associés pour la Recherche Européens (RARE), Réseaux IP Européens*

RIPE NCC

The RIPE Network Coordination Center supports *RIPE* activities which cannot be effectively performed by volunteers from the participating organizations. The services include: *network management* database administration, management of the delegated *Internet Registry* functions, documentation storage, and interactive information services. See also: *Network Information Center.*

RJ-11

A standard 4-wire (2-pair) telephone jack/connector. RJ-11 is used on most single-line telephones. The diagram below shows a view looking into a *jack.*

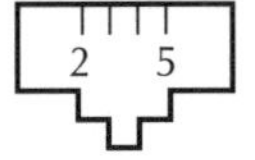

RJ-12

A standard 6-wire (3-pair) telephone jack/connector. RJ-12 is used on most two-line telephones. The diagram below shows a view looking into a *jack.*

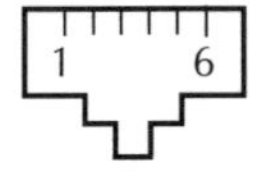

RJ-45

A standard 8-wire (4-pair) jack/connector. RJ-45 is used as a low-cost alternative to standard *twisted-pair* cabling when connecting low-speed devices. The diagram below shows a view looking into a *jack.*

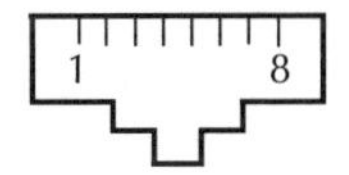

RJE

See: *Remote Job Entry*

rlogin

A protocol/program available with *BSD UNIX* which allows users to login to remote BSD UNIX systems. See also: *remote login, Telnet.*

RO

See: *Receive Only*

ROSE

See: *Remote Operations Service Element*

round-robin

A *time-slot* allocation method in which each user gets a slot in turn. After the last user has had a turn, the next slot goes to the first user and the sequence repeats.

Round-Trip Time (RTT)

A measure of the current delay on a network.

route

The information describing the *path* taken by packets from one node to another. Routes are determined by *routing protocols* or provided by *network administrators* in the form of *static routes.* See also: *asymmetric route.*

routed

Route Daemon. A program which runs under 4.2BSD/4.3BSD *UNIX* systems (and derived operating systems) to propagate routes among machines on a local area network, using the *RIP* protocol. It is often pronounced "route-dee." See also: *gated.*

router

A device which *forwards* traffic between networks. The *routing* decision is based on network-layer information and routing tables, often constructed by *routing protocols.* See also: *bridge, gateway, Exterior Gateway Protocol, Interior Gateway Protocol.*

routing

The process of selecting the correct *interface* and next *hop* for a *packet* being forwarded. In hop-by-hop routing, each *router* in the *path* determines the next hop, without regard to the path already taken. In source routing, the route is determined by the originator of packet, stored in the packet, and executed by the routers in the path. Routing should not be confused with *forwarding*. See also: *Distance-Vector algorithm, link-state algorithm, Exterior Gateway Protocol, Interior Gateway Protocol.*

routing domain

A set of *routers* exchanging routing information within an *Autonomous System*. See also: *Administrative Domain.*

Routing Information Protocol (RIP)

A *routing protocol* used by IP. It is based on a *Distance-Vector algorithm*. RIP is an Internet Standard *IGP* defined in RFC 1058 and extended in RFC 1388. Other forms of RIP are used by the *IPX* and *XNS* protocol suites. See also: *Open Shortest Path First.*

routing loop

A pathological condition in which the *route* for a packet contains the same *router* multiple times. In such a case, the packet circulates between the routers in the loop (i.e., the routers between the two occurrences of the same router) until the *Time-To-Live* for the packet expires. See also: *black hole.*

routing protocol

A protocol used by *routers* to exchange network *topology* information. This exchange of information allows routers determine the topology of a network so that they may select the optimal *path* for *messages* travelling through that network. See also: *Distance-Vector algorithm, link-state algorithm, IS-IS, RIP, RTMP, OSPF.*

Routing Table Management Protocol

A *routing protocol* used by *Appletalk*. It allows *routers* to determine the *path* between two *sockets* using a *Distance-Vector algorithm.*

RPC

See: *Remote Procedure Call*

RS-232

The most common, standard *interface* used to connect *DTEs* to *modems.* It uses a *DB-25* connector, although the *DB-9* version has become popular on *PCs* which have limited space for connectors. The following table gives the pinout for the DB-9 connector and lists the corresponding pin number for a DB-25. The extra signals available on the DB-25 are not required for most *asynchronous*, *serial interface* connections. Each of the indicated signals are described in this glossary.

DB-9	DB-25	Signal name	To DTE	To DCE
1	8	Carrier Detect (CD)	X	
2	3	receive data	X	
3	2	transmit data	X	
4	20	Data Terminal Ready (DTR)		X
5	7	signal ground		
6	6	Data Set Ready (DSR)	X	
7	4	Request To Send (RTS)		X
8	5	Clear To Send (CTS)	X	
9	22	Ring Indicator (RI)	X	

RS-422

A modernized electrical specification for the basic *RS-232* interface. It allows for higher speeds and longer cable runs. See also: *RS-449*.

RS-449

An expanded *RS-232* interface using a *DB-37* connector. It allows for more *modem* control signals and is used with the *RS-422* electrical specification.

RSA

The Rivest, Shamir, Adleman *public key encryption* standard used for *authentication* and data privacy. See also: *Data Encryption Standard*.

RSN

Real Soon Now

RTFM

See: *Read the F*cking Manual*

RTMP

See: *Routing Table Maintenance Protocol*

RTS

See: *Request To Send*

RTSE

See: *Reliable Transfer Service Element*

RTT

See: *Round-Trip Time*

RZ

See: *Return to Zero*

S

SAA

See: *Systems Application Architecture*

SACK

See: *Selective Acknowledgment*

SAP

See: *Service Access Point, Service Advertising Protocol*

satellite communication

The use of one or more satellites, usually in *geosynchronous orbit*, to relay transmissions between *earth stations*. See also: *down-link, up-link, transponder.*

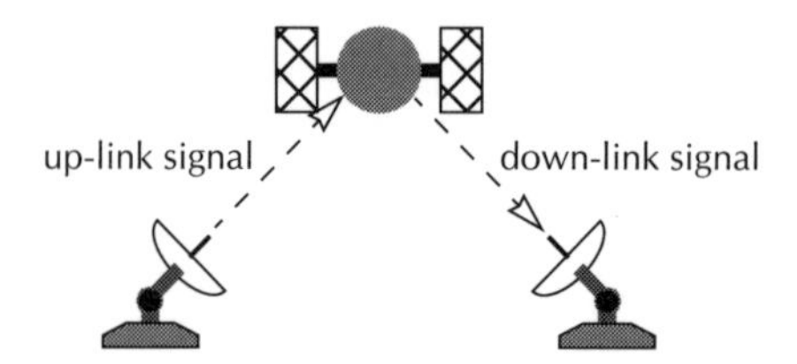

scattering

The loss of *light signal* strength in a *fiber-optic* cable caused by variations in the material composing the fiber. Also, the diffusion of an *open-air transmission* caused by varying densities of the air.

SCP

See: *Session Control Protocol*

SCS
See: *SNA Character String*

SCSI
See: *Small Computer Standard Interface*

SDLC
See: *Synchronous Data Link Control*

SDU
See: *Service Data Unit*

segment (hardware)
The physical *medium* which connects *active devices* on a network. For example, the *Twisted-Pair* cable in a *point-to-point link*, or the *coaxial cable* in an *802.3 LAN*.

segment (software)
A block of *TCP* data. Although TCP is a *stream* protocol, it may break an *application's messages* into segments to attempt to prevent *IP* from *fragmenting* the message. Segmentation is more efficient that *fragmentation* because fragments must be *reassembled* and segments can be delivered as they are received.

selection
The process by which one *SNA* station contacts another to send it a *message*. See also: *polling*.

selective acknowledgment
An extension to the basic *sliding window* algorithm. It allows the receiver to *acknowledge* data received out of sequence (i.e., before the expected data), but still within the window. See also: *flow control*.

selective-repeat
A form of *Automatic Repeat Request*. The receiver indicates to the transmitter that a packet has been damaged or lost, and the transmitter *retransmits* only that packet. It is the most complex *ARQ* to implement, but it makes the most efficient use of available *bandwidth* and produces the highest *throughput*. See also: *go-back-N, stop-and-wait*.

selector

An identifier which is used by a layer in the *OSI protocol stack* to distinguish among the multiple *SAPs* it provides to the layer above. See also: *port.*

Sequenced Packet Exchange (SPX)

Novel's *transport-layer* protocol. It operates over *IPX* and provides reliable, *connection-oriented* service. It is not a *stream-oriented* service; therefore, it preserves the packet boundaries for an *application's* messages. It uses a *go-back-N* algorithm to provide guaranteed delivery.

Sequenced Packet Protocol (SPP)

An *XNS transport-layer* protocol which provides guaranteed, ordered delivery of packets.

sequencing

See: *fragmentation*

serial interface

A network *interface* which transmits each of the bits in a character sequentially. See also: *parallel interface, Universal Asynchronous Receiver/Transmitter.*

Serial Line IP (SLIP)

A protocol used to run *IP* over *asynchronous, point-to-point* lines. It has being succeeded by *PPP.*

server

A provider of a resource or resources (e.g., *file servers, print servers*). See also: *client, Domain Name System, Network File System.*

service

An *application* or function provided by a *host* or a *server. File transfer, remote login,* and *name resolution* are examples of services available on most networks.

Service Access Point (SAP)

The interface between a *layer* in the *OSI protocol stack* and the layer above. Generally, "SAP" is preceded by a letter denoting the layer providing the service (e.g., *network-layer* services are NSAPs). Well known services are associated with well known SAP numbers. See also: *port.*

Service Advertising Protocol (SAP)

An *IPX* protocol which *routers* use to determine how to reach network nodes which provide special services (e.g., *file servers*). It is similar to the *RIP* routing protocol, except that it exchanges information about services instead or networks.

Service Data Unit (SDU)

The *message* to be passed to a *peer* layer on a remote node. See also: *Interface Data Unit, Protocol Control Information, Protocol Data Unit.*

serving area

In broadcasting, the geographic area in which the signal strength of the broadcast is at or above a receivable minimum. In *telephony*, the geographic area handled by a *Central Office.*

session

A logical *session-layer* connection between users on two network *nodes.*

Session Control Protocol (SCP)

A *DECnet session-layer* protocol which provides access control, *process* administration, and *name resolution.*

session layer

The fifth layer of the *OSI reference model.* This minimal layer is responsible for managing *sessions* and transferring data over them. Basically, the session layer adds value to the *transport layer* in the form of *dialog management* and error recovery.

SGML

See: *Standard Generalized Markup Language*

SGMP

See: *Simple Gateway Monitoring Protocol*

SHF

See: *Super High Frequency*

short frame

A *frame* with a length less than the minimum allowable length for the *medium* over which the frame was transmitted. For example, *Ethernet* has a minimum frame length of 64 bytes.

shielding

A metallic sheath surrounding a transmission *medium*, or enclosure for an electronic device, designed to minimize the *Electromagnetic Interference* and *Radio Frequency Interference* leaking from a cable or device.

SIG

Special Interest Group

signal converter

A device which accepts an incoming signal and transforms it for output, typically for transmission onto another *medium.*

signal element

A discrete waveform representing a binary digit in an *analog* transmission.

Signal-to-Noise Ratio (SNR)

The relationship of the strength (magnitude) of a signal to the amount of background noise on the transmission *medium.*

signature

The three or four line message at the bottom of an *email* message or a *Usenet* article which identifies the sender. Large signatures (over five lines) are generally frowned upon. See also: *Electronic Mail.*

Simple Class

See: *Transport Class 0*

Simple Gateway Monitoring Protocol (SGMP)

The *Internet* predecessor to *SNMP.*

Simple Mail Transfer Protocol (SMTP)

A protocol used to transfer *Electronic Mail* between computers. It is a server to server protocol, so other protocols are used to access the messages. See also: *Post Office Protocol, RFC 822.*

Simple Network Management Protocol (SNMP)

The *Internet* Standard *protocol* developed to manage *nodes* and devices on an *IP* network. It is currently possible to manage *hosts, routers, wiring hubs,* toasters, jukeboxes, etc. See also: *Management Information Base.*

simplex

The operation of a communication *channel* which permits data to be transmitted in only one direction. There is no capacity for transmission in both directions. See also: *Full Duplex, Half Duplex.*

Simultaneous Peripheral Operation On-Line (SPOOL)

The ability for multiple users to share a *peripheral device* by saving the output from each user on disk, then sending each output sequentially to the peripheral device.

single mode

A *fiber-optic* cable which transmits *light signals* of only a single frequency and phase. Typically, a *laser* is used to generate the signals. See also: *multi-mode, optical fiber.*

sink

A receiver of data. See also: source.

SITA

See: *Société Internationale de Télécommunication Aéronautique*

skewing

The time delay (offset) between any two signals.

sliding window

A *flow control* mechanism wherein the size of the window is equal to the maximum number of data units (*characters* or *messages*, depending on the protocol) which may be transmitted before an *acknowledgment* is received.

SLIP

See: *Serial Line IP*

slotted Aloha

A variation on the basic *Aloha* protocol in which stations are synchronized to a central clock and may only transmit data at the beginning of a *time slot.* The maximum length of a packet is determined by the length of a slot. This variation ensures that if a transmission collision occurs, it will be a total collision, thus eliminating the possibility of one node's transmission being damaged when it is nearly complete.

Small Computer Standard Interface (SCSI)
A disk-drive to disk-controller *interface* standard. It is typically used on workstations, but is available for *PCs*. It is often pronounced "scuzzy."

SMDS
See: *Switched Multimegabit Data Service*

SMFA
See: *Specific Management Functional Area*

SMI
See: *Structure of Management Information*

SMTP
See: *Simple Mail Transfer Protocol*

SNA
See: *Systems Network Architecture*

SNA Character String (SCS)
A data transmission format consisting of *EBCDIC* control characters, possibly mixed with user data, which is carried within an *SNA* request or response *message.*

SNA Delivery System (SNADS)
An *SNA application* providing *store-and-forward* delivery of documents.

SNADS
See: *SNA Delivery System*

SNAP
See: *Subnetwork Access Protocol*

SNDCF
See: *Subnetwork Dependent Convergence Facility*

snail mail
A pejorative term referring to the United States postal service.

SNMP
See: *Simple Network Management Protocol*

SNR

See: *Signal-to-Noise Ratio*

Société Internationale de Télécommunication Aéronautique (SITA)

The international data communication network used by many airlines.

socket

The combination of *network address, host address* and service identifier (*port*) which uniquely identifies one end of a network connection.

sockets interface

The *UNIX* interface between *application* programs and the networking *protocol stack*. While the protocol stack is usually *TCP/IP*, the sockets interface provides for multiple protocol stacks operating in parallel. See also: *Application Program Interface, POSIX.*

SOH

See: *Start Of Header*

SOM

See: *Start Of Message*

SONET

See: *Synchronous Optical Network*

source

A transmitter of data. See also: *sink*.

source quench

A *flow control* command, sent by a receiver to a transmitter, which indicates that the transmitter should pause momentarily. The exact semantics of a source quench depend on the *protocol* in use. See also: *congestion control.*

source route

A *route* which is determined by the originator of a *datagram*. That route is stored into the datagram so that the *routers* along the *path* can execute it.

source routing

See: *routing*

Southwestern Bell

The *Regional Bell Operating Company* which services the southwestern region of the United States.

SPAG

See: *Standards Promotion and Application Group*

Specific Management Functional Area (SMFA)

One of *OSI's network management* standards which defines individual network management *services.* See also: *Structure of Management Information, CMIP, CMIS.*

spectrum

The entire range of *Radio Frequencies.* See also: *electromagnetic spectrum.*

spectrum allocation

The assignment of *Radio Frequencies,* to specific uses and users, for the purpose of preventing *interference* between unrelated transmissions. In the United States, the allocation is done by the *Federal Communications Commission.*

SPOOL

See: *Simultaneous Peripheral Operation On-Line*

SPP

See: *Sequenced Packet Protocol*

SPX

See: *Sequenced Packet Exchange*

SQL

See: *Structured Query Language*

SSAP

Source Service Access Point (see: *Service Access Point*)

SSCP

See: *System Services Control Point*

stack

A temporary data storage construct onto which *messages* may be placed and from which they are removed. Messages may only be placed on, and removed from, the top of the stack. See also: *LIFO, queue.*

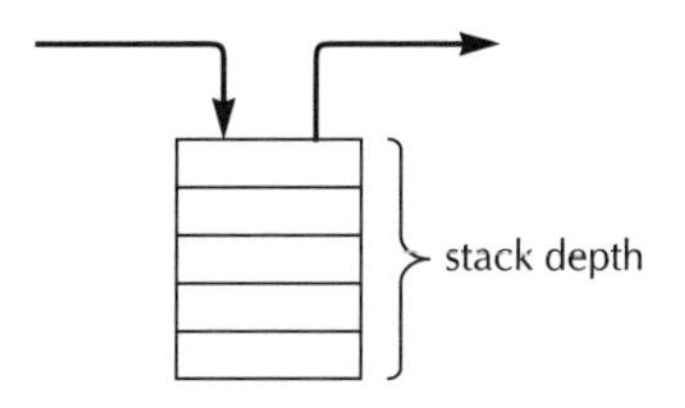

standard

A hardware or software specification which has been generally accepted and implemented, in the case of a de facto standard, or approved by an appropriate standards making organization (e.g., *ANSI, IETF, ISO*), in the case of a du jure standard. See also: *implementors' agreement.*

Standard Generalized Markup Language (SGML)

An *ISO* standard text representation language.

Standards Promotion and Application Group (SPAG)

A group of European *OSI* manufacturers which chose subsets of OSI options and publishes them in a *Guide to the Use of Standards.*

star network

A network *topology* in which multiple network nodes are connected through a single, central node. The central node is a device which manages the network. This topology suffers from dependence on a central node, the failure of which would bring down the network. See also: *star-wired ring network, lobe.*

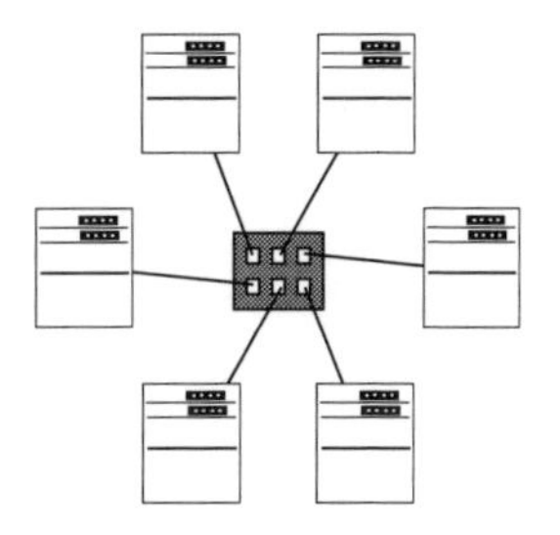

star-wired ring network

A network *topology* in which multiple network nodes, comprising a *star network*, are connected to a *ring network*. That is, the *wiring hubs* are connected in a ring topology and the nodes connect to the wiring hubs. See also: *lobe*.

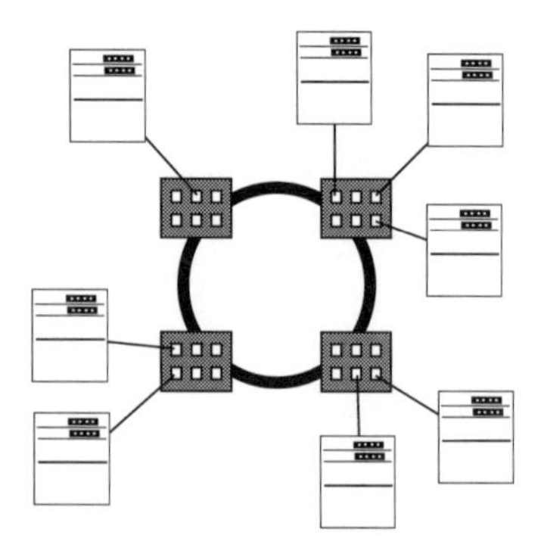

StarLAN

See: *1base5*

start bit

The first bit of each character transmitted over an *asynchronous* communication circuit. It prepares the receiving device for reception of the character. See also: *stop bit*.

Start Of Header (SOH)

A *control character* which marks the beginning of control information in a *character-oriented* data stream.

Start Of Message (SOM)

A *control character* which marks indicates the beginning of a list of addresses for nodes which are to receive this *message*. See also: *End Of Message*.

Start of Text (STX)

A *control character* which indicates the end of control information and the beginning of a *message's* text. See also: *End of Text*.

static route

A *route* configured by a network administrator. See also: *default route, Dynamic Adaptive Routing routing protocol.*

station

Any device which can receive and transmit *messages* on a network. See also: *node, End System, host, Intermediate System, router, gateway.*

STD

A subseries of *RFCs* which specify *Internet* Standards. The official list of Internet Standards is in STD 1. See also: *For Your Information, Request For Comments.*

stop-and-wait

A form of *Automatic Repeat Request.* It requires the transmitter and receiver to operate in *lockstep.* Guaranteed, ordered delivery of data is ensured because there is never more than one packet outstanding at any time. It is a simple but inefficient algorithm because the waiting for a *acknowledgment* wastes *bandwidth.* See also: *go-back-N, selective-repeat.*

stop bit

The last bit of each character transmitted over an *asynchronous* communication circuit. See also: *start bit.*

store-and-forward

An network operational mode in which *messages* are received in their entirety before being transmitted to the destination, or to the next *hop* in the *path* to the destination. See also: *gateway, router.*

stream

A succession of data characters which must be kept in sequence but which are not grouped into blocks.

stream-oriented

A type of service which allows data to be sent in a continuous *stream*, as opposed to fixed-size blocks. This type of service guarantees reliable, ordered, non-duplicated data delivery; however, it explicitly does not preserve *message* boundaries when *application* data is blocked.

streams

A *UNIX* V.3 framework for network communications. It is based on a stacked module concept, wherein each module performs a network service (e.g., *IP, TCP*) and can communicate with modules immediately above and below it.

Structure of Management Information (SMI)

The rules used to define the objects which can be accessed via a *network management* protocol. See also: *Management Information Base, CMIP, CMIS, SNMP.*

Structured Query Language (SQL)

The international standard language used to define and access *relational databases.*

stub network

A *network* which only carries packets to and from local hosts. Even if it has *paths* to more than one other network, it does not carry traffic for other networks. See also: *backbone, transit network.*

STX

See: *Start of Text*

subchannel

A portion of a channel. *T1*, for example, has 24 64Kb/s subchannels.

Subnetwork Dependent Convergence Facility (SNDCF)

An *OSI* service sublayer which allows *CLNP* to operate over an *X.25* network.

subnet

A portion of a *network*, which may be a physically independent network segment, which shares a *network address* with other portions of the network and is distinguished by a *subnet number.* A subnet is to a network what a network is to an *internet.*

subnet address

The *subnet* portion of an *IP address.* In a subnetted network, the host portion of an IP address is split into a subnet portion and a host portion using an *address mask.* See also: *network address, host address.*

subnet broadcast

A *broadcast* packet which is received by all nodes on the specified *subnet.* In *IP,* for example, a packet addresses to 132.245.33.0 would be delivered to all nodes on subnet 132.245.33.0. Where necessary, routers will forward the packet. See also: *directed broadcast, network broadcast, multicast.*

subnet mask

See: *address mask*

subnet number

See: *subnet address*

Subnetwork Access Protocol (SNAP)

A field appended to *802.2* packets, with source and destination *SAPs* of AAh, used to distinguish between non-*OSI* protocol families (e.g., *AppleTalk, IP, IPX*).

Super High Frequency (SHF)

A portion of the *electromagnetic spectrum* (microwave) with frequencies ranging from 3GHz to 30GHz.

SVC

See: *Switched Virtual Circuit*

switched line

A communication circuit for which the physical *path* is established at call setup. After the circuit has been dropped, a later setup to the same destination may follow a different path.

Switched Multimegabit Data Service (SMDS)

A high-speed *datagram*-based *Public Data Network* service developed by Bellcore and expected to be widely used by telephone companies as the basis for their data networks. See also: *Metropolitan Area Network.*

Switched Virtual Circuit (SVC)

An *X.25 dial-up* connection which remains dedicated to the end points of the connection until the connection is terminated. See also: *Permanent Virtual Circuit.*

symbol

See: *signal element*

SYN

A *control character* used to synchronize the transmission timing between a transmitter and a receiver.

synchronous

A transmission method in which bits are sent at a fixed rate. The transmitter and receiver are kept in synchronization by a clocking signal residing in one of the nodes on the synchronous circuit, or a clocking signal generated by the provider of the circuit. Start and stop bits are not necessary because of the clocking. See also: *asynchronous.*

Synchronous Data Link Control (SDLC)

IBM's version of *HDLC.* It is the primary *SNA datalink-layer* protocol.

Synchronous Optical Network (SONET)

A *datalink-layer* protocol for *T3*, *fiber-optic* networks.

System Services Control Point (SSCP)

An *SNA* network entity which manages network configuration, provides directory services, and maintains network address and *mapping* tables.

Systems Application Architecture (SAA)

The *architecture* for IBM's software communications products. It specifies the *Application Program Interface* and communications support for IBM's *Operating Systems.*

Systems Network Architecture (SNA)

A proprietary networking *architecture* used by IBM and IBM-compatible mainframe computers.

T

T1

A communications carrier facility used to transmit *DS-1* formatted *digital* signals at 1.544 Mb/s.

T3

A communications carrier facility used to transmit *DS-3* formatted *digital* signals at 44.746 Mb/s.

TA

See: *Terminal Adapter*

TAC

See: *Terminal Access Controller*

TACACS

See: *Terminal Access Controller Access Control System*

talk

A protocol which allows two people on remote computers to communicate in a *real-time, interactive* fashion. See also: *Internet Relay Chat.*

tap

An access point into a communication circuit or onto a communication *medium.*

TASI

See: *Time-Assigned Speech Interpolation*

TAT-8

The eighth transatlantic communications cable, and the first to use *single-mode fiber-optic* technology.

TCAM

See: *Telecommunications Access Method*

TCP

See: *Transmission Control Protocol*

TCP/IP Protocol Suite

This is a common shorthand which refers to the suite of *transport-layer* and *application-layer* protocols which operate over the *Internet Protocol.*

TDM

See: *Time Division Multiplexing*

TDMA

See: *Time Division Multiple Access*

Technical and Office Protocols (TOP)

A subset of *OSI* standards, originated by Boeing, designed to maximize interoperability in areas where plain OSI standards are ambiguous or allow excessive options. See also: *Government OSI Profile, Manufacturing Automation Protocol.*

Telco

Telephone Central Office

telecommunications

A generic term for voice and data communication, typically using the telephone system as the network. See also: networking, telephony.

Telecommunications Access Method (TCAM)

An IBM communication software package which can support pre-*SNA* and *SNA* protocols. Although TCAM is being replaced by *VTAM*, it is still widely deployed.

TELENET

A *packet-switched, Public Data Network* using the *CCITT X.25* protocols. It should not be confused with the *Telnet* application protocol.

telephony

A generic term applied to the hardware and protocols used by the telephone companies.

teleprinter exchange (Telex)

An international, switched message-exchange service which uses *Baudot* encoding.

teletypewriter

A generic term for a data communication terminal which outputs to paper. See also: *Cathode Ray Tube.*

Telex

See: *teleprinter exchange*

Telnet

The *Internet* Standard protocol for remote terminal connection service. It has been extended with options defined in many RFCs.

terminal

A peripheral device which allows users to *interactively* access a computer and, through it, a data communication network. See also: *Cathode Ray Tube, Teletypewriter.*

Terminal Access Controller Access Control System (TACACS)

The *authentication* protocol and authentication servers used by *TIP*s to validate a user for *dial-up* access.

Terminal Adapter (TA)

A device which allows non-*ISDN* devices to be connected to an ISDN network.

terminal emulator

A program which allows a computer to *emulate* a *terminal.* The workstation or *PC* running the emulator thus appears as a terminal to the remote host.

terminal node

An *SNA* peripheral node which is not user programmable.

terminal server

A device which connects many terminals to a *LAN* through one network connection. A terminal server can also connect many network users to its *asynchronous* ports for dial-out capabilities and printer access.

terminated line

A telephone circuit with a resistance (load) at the remote end which is equal to the *impedance* of the line. The effect of the load is to eliminate signal reflection (*echo*).

terminator

The 75Ω load attached to both ends of an *10Base2* or *10Base5* cable. The load prevents signal *echos*, which would otherwise be considered *collisions.* See also: *impedance, coaxial cable.*

terrestrial

Describing a communications *path* which does not use *satellite communications.* See also: *land line.*

TFTP

See: *Trivial File Transfer Protocol*

thicknet

See: *10base5*

thinnet

See: *10base2*

Three Letter Acronym (TLA)

A tribute to the use of acronyms in the computing industry. See also: *Extended Four Letter Acronym.*

three-way handshake

A synchronization procedure used by many *protocols* for connection establishment and take-down, and database transactions. It requires three *messages* to pass between the two communicating *processes* and is more reliable than a *two-way handshake.* See also: *handshake.*

throughput

The amount of data which passes by a point on the network within a certain period of time. See also: *carrying capacity, transmission latency.*

tickle

An AppleTalk *keep-alive* packet sent every 20 seconds, regardless of user data transmissions.

tie line

A dedicated communication circuit, provided by *common carriers*, which links two points without using the switched telephone network.

Time-Assigned Speech Interpolation (TASI)

A technique wherein the pauses between words are used to carry additional voice communications. In effect, multiple conversations are *multiplexed* over a single channel.

Time Division Multiple Access (TDMA)

A satellite access technique wherein several *earth stations* use, in *round-robin* fashion, the total transponder power and available *bandwidth*. See also: *Time Division Multiplexing.*

Time Division Multiplexing (TDM)

A technique for sharing a single communications *channel* among multiple users. Each user is allocated a *time-slot* during which it may transmit data, thus the data on the channel is interleaved. The *interleaving* may be by bit or by character. See also: *Frequency Division Multiplexing.*

time-out

The expiration of a specified period of time. A time-out is generally used to improve data throughput by generating *retransmissions* of unacknowledged data. A time-out may also be used to terminate a connection which has been *idle* for a specified period of time.

time-slot

An allocation of time for a resource being *time-shared*, or a communication channel being *Time Division Multiplexed.*

Time To Live (TTL)

A field in an *IP datagram header* which indicates how long this packet should be allowed to survive before being discarded. It is primarily used as a *hop* count limit. See also: *Maximum Packet Lifetime.*

timesharing

The *interleaving* of multiple users' requests onto a shared resource.

TIP

An *IMP* which supported *dial-up*, terminal access to the *ARPANET.* See also: *Terminal Access Controller Access Control System.*

tip

An *application* which is used to establish a *Full Duplex* connection to a remote host using a *TTY* port to which a *modem* has been attached.

TLA

See: *Three Letter Acronym*

TLAP

See: *TokenTalk Link Access Protocol*

TN3270

A variant of the *Telnet* application which allows a user to attach to IBM mainframes and use the mainframe as if you had a 3270, or similar, terminal. See also: terminal emulator.

token

A unique bit pattern which all stations on the *LAN* recognize as a "permission to transmit" indicator. See also: *CSMA/CD, token passing.*

token bus

A type of *Local Area Network* with the nodes connected to a common cable (bus). While *topologically* similar to *Ethernet* and *802.3*, token bus uses a *CSMA/CA* protocol rather than a *CSMA/CD* protocol. See also: *802.4, token ring.*

token passing

An access method in which a *token* is used to control access to the network. Only the station with the token may transmit, and, after transmitting, it must pass the token to the next station on the network. See also: *CSMA/CA, token bus, token ring.*

token ring

A type of *Local Area Network* with stations wired into a *ring network.* Each station constantly passes a *token* on to the next; whichever station has the token may send a *message.* Often, *Token Ring* is used to refer to the *IEEE 802.5* token ring standard, which is the most common type of token ring. See also: *802.x, token bus, CSMA/CA.*

Token Rotation Time (TRT)

The amount of time it takes a *token* to go around an *802.5* ring. The time varies with the size of the ring and the number of *stations* attached.

TokenTalk Link Access Protocol (TLAP)

The *AppleTalk datalink-layer* protocol used over *802.5* networks. See also: *AppleTalk Remote Access Protocol, EtherTalk Link Access Protocol, LocalTalk Link Access Protocol.*

TOP

See: *Technical and Office Protocols*

topology

The relationship between the nodes in the network and the arrangement of the network cabling. Typical simple topologies include: bus, ring, and star. See also: *fully connected network, mesh network, ring network, star network.*

TP

See: *Twisted-Pair*

TP0

See: *Transport Class 0*

TP4

See: *Transport Class 4*

trailer

Control information, usually a *checksum, CRC* or *FCS*, appended to a packet. See also: *header.*

transceiver

See: *transmitter-receiver*

transceiver cable

A cable which connects a node to a *transceiver* on a *Local Area Network.* A transceiver cable should not be confused with a *drop cable.*

transit network

A *network* which passes traffic between other networks in addition to carrying traffic for its own nodes. It must have connections to at least two other networks. See also: *backbone, stub network.*

transmission

The sending of a signal or *message* by any means (e.g., fiber, radio, wire).

Transmission Control Protocol (TCP)

An *Internet* Standard *transport-layer* protocol. It operates over *IP* and is *connection-oriented* and *stream-oriented.* It uses a *sliding-window* algorithm to guarantee delivery. See also: *Transport Class 4, User Datagram Protocol.*

transmission latency

The time it takes for a *message* to travel between two devices on a network. See also: *propagation delay, throughput.*

transmitter-receiver (transceiver)

The generic name for a device which is capable of transmitting and receiving. In a *LAN* context, it is the device which connects a node to the network. See also: *transceiver cable.*

transparency

A characteristic of a communication medium or *protocol* in which operation of that medium or protocol is hidden from the user.

transponder

The circuitry in a satellite which receives an *up-link* signal from one *earth station*, transforms it into another frequency, and transmits it as a *down-link* signal to another earth station.

Transport Class 0 (TP0)

An *OSI transport-layer* protocol, often referred to as Simple Class. It is a *connectionless* protocol which adds a level of reliability and *multiplexing* to the network protocol. See also: *User Datagram Protocol, Transport Class 4.*

Transport Class 4 (TP4)

An *OSI transport-layer* protocol, often referred to as Error Detection and Recovery Class. It is a *connection-oriented protocol* which guarantees orderly, reliable delivery of data. See also: *Transmission Control Protocol, Transport Class 0.*

transport layer

The fourth layer of the *OSI reference model.* This layer is responsible for *end-to-end*, as opposed to *hop-to-hop* delivery of data. *TCP, UDP, TP4* and *TP0* are examples of transport-layer protocols.

trellis coding

A type of *Forward Error Correction* available in some high-speed *modems.* It compares the *encoding* of a received *signal element* to the encoding of the previously received signal element to determine if an error has occurred.

Trivial File Transfer Protocol (TFTP)

A *UDP*-based *file transfer* protocol which operates by transmitting 512-byte blocks of data in a *lockstep* fashion. It is easy to implement, but offers no file manipulation other than read and write. See also: *File Transfer Protocol.*

Trojan Horse

A computer program which carries within itself a means to allow the creator of the program to access the system using it. See also: *virus, worm, cracker.*

TRT

See: *Token Rotation Time*

trunk

A dedicated telephone circuit which interconnects telephone switching centers and *Central Offices.*

trunk group

Multiple trunks connecting the same two switching centers which may be accessed with a single trunk number.

TTFN

Ta-Ta For Now

TTL

See: *Time to Live*

TTY

The term applied to a generic input/output device or device *interface.* See also: *Cathode Ray Tube, teletypewriter.*

tunnelling

The *encapsulation* of *protocol* A within protocol B, such that A treats B as though it were a *datalink layer.* Tunnelling is used to get data between *Administrative Domains* which use a protocol that is not supported by the *internet* connecting those domains. See also: *backbone, transit network.*

turnaround

The change in transmission direction on a *Half Duplex* circuit.

turnaround time

The time it takes a *Half Duplex* circuit to change transmission direction. Also, the time it takes telephone circuit echo suppressors to change state.

twinaxial cable (twinax)

A variant of *coaxial cable* which contains two central conductors instead of one.

Twisted-Pair (TP)

A type of cable in which pairs of conductors are twisted together to produce certain electrical properties. See also: *Unshielded Twisted-Pair.*

two-way handshake

A synchronization procedure used by many protocols for connection establishment and take-down, and database transactions. It requires two *messages* to pass between the two communicating *processes.* It is less reliable than a *three-way handshake.* See also: *handshake.*

Type Of Service (TOS)

A field in *IP datagram headers* which may be used to request special treatment for a datagram. The currently defined types of service are: low delay, high throughput, and high reliability. See also: *Quality Of Service.*

U

UA

See: *User Agent*

UART

See: *Universal Asynchronous Receiver/Transmitter*

UDP

See: *User Datagram Protocol*

UHF

See: *Ultra High Frequency*

ULF

See: *Ultra Low Frequency*

Ultra High Frequency (UHF)

A portion of the *electromagnetic spectrum* with frequencies ranging from 300MHz to 3GHz. Television channels 14 through 83 and cellular radio fall within this frequency range.

Ultra Low Frequency (ULF)

A portion of the *electromagnetic spectrum* with frequencies below 30Hz. Electrical power is transmitted over power lines at these frequencies.

ULTRIX

Digital Equipment Corporation's version of *UNIX.*

underrun

A condition which occurs when a *serial interface* driver runs out of data to send before the logical end of the *frame* is reached. In other words, it was transmitting faster than the device could supply data. See also: *overrun.*

Unified Network Management Architecture (UNMA)

AT&T's *architecture* for *network management.*

Universal Asynchronous Receiver/Transmitter (UART)

An *Integrated Circuit* which takes a character of data (eight bits in parallel), and transmits each bit serially over an *asynchronous* communication channel. It also accepts asynchronous data and provides a character to the device in which the UART is installed.

Universal Resource Identifier (URI)

A globally unique name for a document which is accessible via the *World Wide Web* protocol. *URLs* are also a form of URI.

Universal Resource Locator (URL)

A standard format for names which describe the physical addresses of documents accessible via the *World Wide Web* protocol. See also: *Universal Resource Identifier.*

Universal Synchronous/Asynchronous Receiver/Transmitter (USART)

An *Integrated Circuit* which combines the functions of a *Universal Synchronous Receiver/Transmitter* and a *Universal Asynchronous Receiver/Transmitter.*

Universal Synchronous Receiver/Transmitter (USRT)

An *Integrated Circuit* which takes a character of data (eight bits in parallel), and transmits each bit serially over an *synchronous* communication channel. It also accepts synchronous data and provides a character to the device in which the USRT is installed.

UNIX

A very widely deployed operating system developed by AT&T. *AIX, BSD, ULTRIX,* and *XENIX* are examples of UNIX-based operating systems.

UNIX-to-UNIX Copy (UUCP)

This was initially a program run under the *UNIX* operating system which allowed one UNIX system to send files to another UNIX system via *dial-up* phone lines. Today, the term is more commonly used to describe the large international network which uses UUCP to pass news and *Electronic Mail.* See also: *Usenet.*

UNMA

See: *Unified Network Management Architecture*

Unshielded Twisted-Pair (UTP)

A form of *Twisted-Pair* which does not have a metallic shield around the cable.

unsolicited response

A *response packet* sent, usually on a periodic basis, for which no *request packet* was received. That is, there are some protocols which send a response packet to reply to a request packet or query packet which is indistinguishable from a periodically broadcast response packet.

up-link

The signal sent from an *earth station* to a satellite. It will usually be transformed into a *down-link* signal. See also: *transponder.*

upstream neighbor

A term indicating the *neighbor* to which a node passes a packet in a ring network. Every node is the upstream neighbor of its *downstream neighbor.*

uptime

An reference to the amount of time a computer has been operational, or a network available, since the last failure or shutdown. See also: *downtime.*

urban legend

A story, which may have started with a grain of truth, that has been embroidered and retold until it has passed into the realm of myth. It is an interesting phenomenon that these stories get spread so far, so fast and so often. Urban legends never die, they just end up on the *Internet*! Some legends that periodically make their rounds include "The Infamous Modem Tax," "Craig Shrugged/Brain Tumor/Get Well Cards," and "The $250 Cookie Recipe."

URI

See: *Universal Resource Identifier*

URL

See: *Universal Resource Locator*

US West

The *Regional Bell Operating Company* which services the western region of the United States.

USART

See: *Universal Synchronous/Asynchronous Receiver/Transmitter*

Usenet

A collection of thousands of topically named newsgroups, the computers which run the protocols, and the people who read and submit Usenet news. Not all *Internet* hosts subscribe to Usenet and not all Usenet hosts are on the Internet. See also: *Network News Transfer Protocol, UNIX-to-UNIX Copy.*

User Agent (UA)

An *OSI* application which creates, submits and takes delivery of *messages* in an *X.400 Message Handling System.*

User Datagram Protocol (UDP)

An *Internet* Standard *transport-layer* protocol. It is a *connectionless* protocol which adds a level of reliability and *multiplexing* to *IP.* See also: *Transport Class 0, Transmission Control Protocol.*

username

The unique identification for a user on a host.

USRT

See: *Universal Synchronous Receiver/Transmitter*

UTC

Coordinated Universal Time (see: Greenwich Mean Time)

UTP

See: *Unshielded Twisted-Pair*

UUCP

See: *UNIX-to-UNIX Copy*

V

V recommendations

A series of *CCITT* standards covering *modem* operation. See also: V.35, V.42bis.

V.35

A hardware specification for *synchronous,* high-speed, *serial interfaces.* The diagram below shows the male end of a V.35 connector, looking at the pins. See also: *RS-232, RS-449, Clear To Send, Data Set Ready, Data Terminal Ready, Request To Send.*

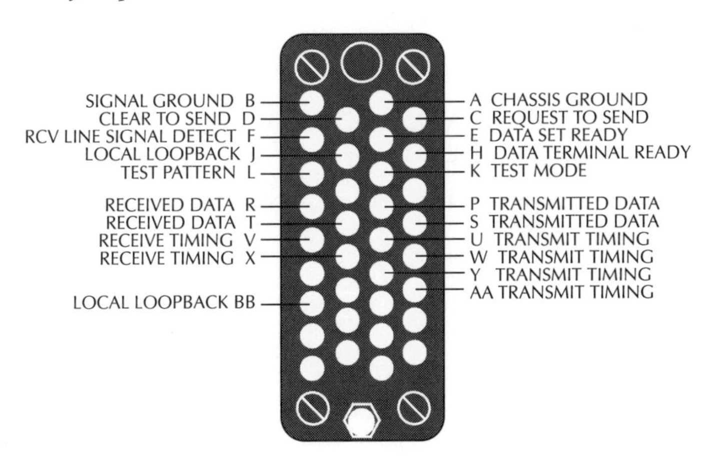

V.42bis

A *CCITT* data *compression* specification for use on *asynchronous* communication circuits. It is frequently implemented in a *modem*'s hardware.

Value-Added Network (VAN)
A communication network which provides services not essential to the basic functioning of such a network.

VAN
See: *Value-Added Network*

Variable Quantizing Level (VQL)
A technique for *encoding* voice, which is a form of *analog* signal, for *digital* transmission at 32Kb/s.

VAX
Virtual Address Extension is a series of computers produced by Digital Equipment Corporation. The plural of VAX is VAXEN.

Very High Frequency (VHF)
A portion of the *electromagnetic spectrum* with frequencies ranging from 30MHz to 300MHz. Television channels 2 through 13 and FM radio fall within this frequency range.

Very Low Frequency (VLF)
A portion of the *electromagnetic spectrum* with frequencies ranging from 3KHz to 30KHz. This frequency is radiated by *CRTs* as unwanted *Radio Frequency Interference.*

Very Small Aperture Terminal (VSAT)
An *earth station* with a small diameter dish. Such stations generally operate in the *Ku band.*

VF
See: *Voice Frequency*

VHF
See: *Very High Frequency*

video teleconferencing
A *real-time,* two-way communication which carries voice and *digitized* video among the facilities participating in the conference. The quality of the video depends on the available *bandwidth.*

VINES
See: *Virtual Network System*

virtual circuit

A network service which provides *connection-oriented* service regardless of the underlying network structure. See also: *connection-oriented.*

Virtual Network System (VINES)

Banyan Systems's Network Operating System.

Virtual Telecommunications Access Method (VTAM)

An IBM communications software product which manages *SNA* communications and *SDLC* communication links.

Virtual Terminal Protocol (VTP)

The *OSI remote login* application protocol. It allows a terminal to connect to a remote system regardless of the terminal type or its characteristics.

virus

A program which replicates itself on computer system by incorporating itself into other programs which are shared among other computer systems. See also: *Trojan Horse, worm, cracker.*

VLF

See: *Very Low Frequency*

Voice Frequency (VF)

A portion of the *electromagnetic spectrum* with frequencies ranging from 300Hz to 3KHz. Most human speech falls into this frequency, so telephone voice-grade channels are optimized to this frequency range.

voice-grade channel

A communication circuit optimized to carry *Voice Frequency* data. Typically, it is not a high quality circuit (i.e., there is background noise) because the human ear is not as sensitive to the noise as a communication device and it is expensive to remove all of the noise.

VSAT

See: *Very Small Aperture Terminal*

VT

Virtual Terminal

VTAM

See: *Virtual Telecommunications Access Method*

VTP

See: *Virtual Terminal Protocol*

W

W3

See: *World Wide Web*

WAIS

See: *Wide Area Information Servers*

WAN

See: *Wide area network*

wart

A pejorative term applied to a *protocol* extension which some protocol designers might consider to be inelegant.

watchdog

Hardware or software which is responsible for ensuring that a device, connection, system or application is still active. In hardware, for example, a watchdog might cause a *router* to reboot if the router appeared to be suffering from a software problem.

WATS

See: *Wide Area Telephone Service*

waveguide

A metallic conduit designed to contain, direct and focus microwave transmissions.

wavelength

The distance between successive peaks, or troughs, of a sine wave (sinusoidal signal).

Wavelength Division Multiplexing (WDM)

A *fiber-optic* transmission technique for sending parallel data over a *multi-mode* fiber. Each of the bits in a character (usually eight) are transmitted at a different wavelength, but simultaneously. Successive characters are sent serially.

well-known port

A *TCP* or *UDP* port number associated with a specific service. For example, *FTP* is *port* 21 and *Telnet* is port 23. Well-known port numbers range from 1 to 1023. See also: ephemeral port.

WG

See: *Working Group*

white pages

The *Internet* supports several databases which contain basic information about users, such as *email addresses*, telephone numbers, and postal addresses. These databases can be searched to get information about particular individuals. Because they serve a function akin to the telephone book, these databases are often referred to as "white pages." See also: *Knowbot, whois, X.500.*

white space

The spaces, tabs, form feeds, line feeds, and carriage returns which appear in a textual message. See also: *printable characters.*

whois

An *Internet* program which allows users to query a database of people and other Internet entities, such as domains, networks, and hosts. The database is maintained by the *InterNIC.* The information about people shows a person's company name, address, phone number and *email address.* See also: *white pages, Knowbot, X.500.*

Wide Area Information Servers (WAIS)

A distributed information service which offers simple natural language input, indexed searching for fast retrieval, and a "relevance feedback" mechanism which allows the results of initial searches to influence future searches. Public domain implementations are available. See also: *archie, Gopher, Prospero.*

Wide Area Network (WAN)

A *network*, usually constructed with *long-haul*, *point-to-point* lines, which covers a large geographic area. See also: *Local Area Network*, *Metropolitan Area Network*.

Wide Area Telephone Service (WATS)

Special rate programs (e.g., 800 numbers) offered by telephone service providers.

wideband

A generic term referring to any communication *channel* with more *bandwidth* than a standard *voice-grade channel*.

WIMP interface

The Window, Icon, Mouse, Pointer *interface* is the basis for most windowing user interfaces. WIMP was originally developed by Xerox *PARC*, but the acronym came much later.

wiring hub

An communications device which provides *taps* for the *lobes* on *star networks* and *star-wired networks*. See also: *concentrator*.

word

A unit of data on a computer. Depending on the CPU, a word may be two, four or eight bytes long. See also: *byte order*.

Workshop for Implementors of OSI (OIW)

The North American regional forum at which *OSI* implementors agreements are created. It is a peer to the *Asia and Oceania Workshop* and the *European Workshop for Open Systems*.

Working Group (WG)

In the *IETF*, a chartered group of volunteers who focus on a specific issue involving the design, operation or management of the *Internet*. See also: *Birds Of a Feather*.

World Wide Web (WWW or W3)

A hypertext-based, distributed information system created by researchers at CERN in Switzerland. Users may create, edit or browse hypertext documents. The *clients* and *servers* are freely available.

worm

A computer program which replicates itself and is self-propagating. Worms, as opposed to viruses, are meant to spawn in network environments. Network worms were first defined by Shoch & Hupp of Xerox in ACM Communications (March 1982). The *Internet* worm of November 1988 is perhaps the most famous; it successfully propagated itself on over 6,000 systems across the Internet. See also: *Trojan Horse, cracker.*

WRT

With Respect To

WWW

See: *World Wide Web*

WYSIWYG

What You See is What You Get

X

X

A series of *CCITT* standards covering data transmission over *Public Data Networks.* See also: *X.21, X.25, X.121.*

X Windows

The *TCP/IP*-based, network-oriented window management system. Network window systems allow a program to use a display on a different computer. The most widely-implemented window system is X11, a component of MIT's Project Athena. See also: *Application Environment Specification.*

X.121

The *CCITT* and *ISO* standard numbering scheme for *Public Data Networks.*

X.21

A data communication specification developed to describe the interface between *DTEs* and *DCEs* over a *synchronous Public Data Network.*

X.25

A data communications interface specification developed to describe how data passes into and out of *Public Data Networks.* It is a *CCITT* and *ISO* approved protocol suite which defines the protocols used in layers 1 through 3 of the *OSI reference model.*

X.400

The *CCITT* and *ISO* standard for *Electronic Mail.* It is widely used in Europe and Canada.

X.500

The *CCITT* and *ISO* standard for electronic directory services. See also: *white pages, Knowbot, whois.*

XbaseY

A generalization for *baseband* transmission *medium* standards nomenclature. X refers to the maximum speed of the link, in megabits per second (Mb/s). Y refers to the maximum length of any segment of the medium (i.e., the maximum distance between repeaters). See also: *1base5, 10base2, 10base5, 10baseT, XbroadY.*

XbroadY

A generalization for *broadband* transmission *medium* standards nomenclature. X refers to the maximum speed of the link, in megabits per second (Mb/s). Y refers to the maximum length of any segment of the medium (i.e., the maximum distance between repeaters). See also: *XbaseY.*

XDR

See: *External Data Representation*

XENIX

A version of *UNIX* which operates on *Personal Computers.*

Xerox Network System (XNS)

A network developed by Xerox corporation. Implementations exist for both 4.3*BSD* derived systems, as well as the Xerox Star computers.

XID

See: *Exchange Identification*

xmodem

A protocol used primarily by *PCs* to reliably transfer data to and from a host, typically over a *dial-up* connection. It has been succeeded by *ymodem.* See also: *file transfer, Kermit*

XNS

See: *Xerox Network System*

XON/XOFF

A *flow control* mechanism (transmit-on/transmit-off) used by *peripheral devices* (primarily *modems* and printers). It allows a device to signal the transmitter to stop sending data pending a request to resume.

Y

ymodem

A protocol used primarily by *PCs* to reliably transfer data to and from a host, typically over a *dial-up* connection. It supersedes *xmodem* and has been superseded by *zmodem*. See also: *file transfer, Kermit*

YP

Yellow Pages (see: *Network Information Services*)

Z

zero code suppression
See: *ones density*

ZIP
See: *Zone Information Protocol*

ZIP bringback time
The amount of time which must expire between the time an *AppleTalk* network is brought down and the time is brought back up with a new *zone* name.

zmodem
A protocol used primarily by *PCs* to reliably transfer data to and from a host, typically over a *dial-up* connection. It supersedes *xmodem* and *ymodem*. See also: *file transfer, Kermit*

zone
A arbitrary group of network nodes in an *Appletalk* network.

Zone Information Protocol (ZIP)
An *Appletalk session-layer* protocol which is used for *mapping network-number* ranges to *zone* names. See also: *Name Binding Protocol.*

APPENDIX A

Networks & Organizations

Advanced Research Projects Agency (ARPA)
Advanced Research Projects Agency Network (ARPANET)
American National Standards Institute (ANSI)
Ameritech
APNIC
Asia and Oceania Workshop
Bell Atlantic
Bell South
Bellcore
Bitnet
Clearinghouse for Networked Information Discovery and Retrieval (CNIDR)
Coalition for Networked Information (CNI)
Comité Consultatif International de Télégraphique et Téléphonique (CCITT)
Computer Emergency Response Team (CERT)
COMSAT
Cooperation for Open Systems Interconnection Networking in Europe (COSINE)
Coordinating Committee for Intercontinental Research Networks (CCIRN)
Corporation for National Research Initiatives (CNRI)
Corporation for Open Systems (COS)
Corporation for Research and Educational Networking (CREN)
DDN NIC
Defense Advanced Research Projects Agency (DARPA)
Defense Data Network (DDN)

Defense Information Systems Agency (DISA)
Electronic Frontier Foundation (EFF)
Electronic Industries Association (EIA)
European Academic and Research Network (EARN)
European Computer Manufacturers Association (ECMA)
European Workshop for Open Systems (EWOS)
FARNET
Federal Communications Commission (FCC)
Federal Networking Council (FNC)
Institute of Electrical and Electronics Engineers (IEEE)
Interagency Interim National Research and Education Network (IINREN)
International Organization for Standardization (ISO)
International Telecommunications Union (ITU)
International Time Bureau (BIH)
Internet Architecture Board (IAB)
Internet Assigned Numbers Authority (IANA)
Internet Engineering Steering Group (IESG)
Internet Engineering Task Force (IETF)
Internet Registry (IR)
Internet Research Steering Group (IRSG)
Internet Research Task Force (IRTF)
Internet Society (ISOC)
InterNIC
Interoperability Technology Association for Information Processing (INTAP)
MILNET
National Cable Television Association (NCTA)
National Exchange Carrier Association (NECA)
National Institute for Standards and Technology (NIST)
National Public Telecomputing Network (NPTN)
National Research and Education Network (NREN)
National Science Foundation (NSF)
National Telecommunications and Information Administration (NTIA)
NSFNET
NYNEX
On-line Computer Library Catalog (OCLC)
Open Software Foundation (OSF)
OSInet
Pacific Bell

Palo Alto Research Center (PARC)
Personal Computer Memory Card Industry Association (PCMCIA)
Promoting Conference for OSI (POSI)
Réseaux Associés pour la Recherche Européens (RARE)
Réseaux IP Européens (RIPE)
RIPE NCC
Société Internationale de Télécommunication Aéronautique (SITA)
Southwest Bell
Standards Promotion and Application Group (SPAG)
TELENET
US West
Workshop for Implementors of OSI (OIW)

APPENDIX B

Security-Related Terms

Acceptable Use Policy (AUP)
Access Control List (ACL)
audit trail
authentication
authorization
Challenge Handshake Authentication Protocol (CHAP)
Computer Emergency Response Team (CERT)
cracker
Data Encryption Key (DEK)
Data Encryption Standard (DES)
decrypt
encrypt
encryption
firewall
Kerberos
Packet Authentication Protocol (PAP)
Privacy Enhanced Mail (PEM)
public key
RSA
Trojan Horse
virus
worm

APPENDIX C

Applications & Protocols

Address Resolution Protocol (ARP)
Advanced Communications Facility (ACF)
Advanced Program-to-Program Communication (APPC)
Aloha
AppleTalk
AppleTalk Address Resolution Protocol (AARP)
AppleTalk Data Stream Protocol (ADSP)
AppleTalk Echo Protocol (AEP)
AppleTalk Filing Protocol (AFP)
AppleTalk Remote Access Protocol (ARAP)
AppleTalk Session Protocol (ASP)
AppleTalk Transaction Protocol (ATP)
archie
Berkeley Internet Name Domain (BIND)
Binary Synchronous Communications (BSC)
BOOTP
Border Gateway Protocol (BGP)
Challenge Handshake Authentication Protocol (CHAP)
Command Terminal Protocol (CTERM)
Common Management Information Protocol (CMIP)
Connection Oriented Network Service (CONS)
Connectionless Network Protocol (CLNP)
Connectionless Transport Protocol (CLTP)
Datagram Delivery Protocol (DDP)
DECmcc
DECnet
DECwindows

Directory Access Protocol (DAP)
Domain Name System (DNS)
Dynamic Host Configuration Protocol (DHCP)
Electronic Data Interchange (EDI)
End System to Intermediate System (ES-IS)
EtherTalk Link Access Protocol (ELAP)
Exterior Gateway Protocol (EGP)
File Transfer, Access, and Management (FTAM)
File Transfer Protocol (FTP)
Gateway Access Protocol (GAP)
Gopher
High-level Data Link Control (HDLC)
Houston Automatic Spooling Program (HASP)
Inter-Domain Routing Protocol (IDRP)
Inter-Domain Policy Routing (IDRP)
Intermediate System to Intermediate System (IS-IS)
Internet Control Message Protocol (ICMP)
Internet Gateway Routing Protocol (IGRP)
Internet Group Multicast Protocol (IGMP)
Internet Message Access Protocol (IMAP)
Internet Protocol (IP)
Internet Relay Chat (IRC)
Internetwork Packet Exchange (IPX)
Kerberos
Kermit
Knowbot
Link Access Procedure (LAP)
Link Access Procedure-Balanced (LAP-B)
Link Access Procedure-D channel (LAP-D)
Link Control Protocol (LCP)
listserv
Local Area Transport (LAT)
LocalTalk Link Access Protocol (LLAP)
Microcom Networking Protocol (MNP)
Multipurpose Internet Mail Extensions (MIME)
Multi-User Dungeon (MUD)
Name Binding Protocol (NBP)
NetView
NetWare

Network File System (NFS)
Network News Transfer Protocol (NNTP)
Network Time Protocol (NTP)
Open Shortest-Path First (OSPF)
Packet Authentication Protocol (PAP)
Packet Internet Groper (PING)
Point-to-Point Protocol (PPP)
Post Office Protocol (POP)
Printer Access Protocol (PAP)
Privacy Enhanced Mail (PEM)
Prospero
Remote File System (RFS)
Reverse Address Resolution Protocol (RARP)
Routing Information Protocol (RIP)
Routing Table Management Protocol (RTMP)
Sequenced Packet Exchange (SPX)
Sequenced Packet Protocol (SPP)
Serial Line IP (SLIP)
Service Advertising Protocol (SAP)
Session Control Protocol (SCP)
Simple Gateway Monitoring Protocol (SGMP)
Simple Mail Transfer Protocol (SMTP)
SNA Delivery System (SNADS)
Subnetwork Access Protocol (SNAP)
Synchronous Data Link Control (SDLC)
Synchronous Optical Network (SONET)
Systems Network Architecture (SNA)
Telnet
tip
TN3270
TokenLink Link Access Protocol (TLAP)
Transmission Control Protocol (TCP)
Transport Class 0 (TP0)
Transport Class 4 (TP4)
Trivial File Transfer Protocol (TFTP)
UNIX-to-UNIX Copy (UUCP)
User Datagram Protocol (UDP)
Virtual Terminal Protocol (VTP)
white pages

whois
Wide Area Information Server (WAIS)
World Wide Web (WWW or W^3)
Zone Information Protocol (ZIP)